The Spokesm

Substandard – Eyewitness rep

Edited by Tony Simpso

Published by Spokesman f
Bertrand Russell Peace Foun...
Ken Coates: Editor 1970 to 2010

Spokesman 129 **2015**

00
00 (Europe)
00 (UK)
00 (UK)
00 (ex UK)

CONTENTS

Cartoon: with grateful thanks to Steve Bell
Cover: HMS Victorious, Faslane, May 2015.

ISSN 1367 7748 Printed by the Russell Press Ltd., Nottingham, UK ISBN 978 0 85124 848 6

A CIP catalogue record for this book is available from the British Library

Published by the
Bertrand Russell Peace Foundation Ltd.,
Russell House
Bulwell Lane
Nottingham NG6 0BT
England
Tel. 0115 9784504
email:
elfeuro@compuserve.com
www.spokesmanbooks.com
www.russfound.org

In Memoriam

MICHAEL BARRATT BROWN

1918 - 2015

Editorial

Substandard – Eyewitness reports Trident

In a 12,000 word report, Able Seaman William McNeilly details his experiences onboard HMS *Victorious*, one of the Royal Navy's four nuclear powered and nuclear armed submarines that sail from their homeport in Faslane, Scotland. McNeilly also probes security at Faslane itself, and finds it badly wanting.

Good practice in health and safety at work requires people to be diligent in observing what is going on around them, and for risk assessments to be prepared and regularly reviewed, and for designated procedures to be followed. The Royal Navy must have a major task in hand as a result of McNeilly's revelations. Defence Ministers have attempted to dismiss the Able Seaman's findings as 'factually incorrect or the result of a misunderstanding', but the United States Navy may take a different view. The United Kingdom leases 58 Trident ballistic missiles from the US Navy under a system of 'mingled asset ownership' (see *Spokesman 127*). What do the American authorities make of AB McNeilly's report, and its challenging comments about procedures for installing missile inverters onboard US Ohio Class Trident submarines, which seem to be much more stringent than on UK boats?

The words 'factual' and 'courageous' come to mind on reading AB McNeilly's account of his personal experiences at Faslane and his extended patrol on HMS *Victorious*. For the first time in some 30 years, we have a submariner's eyewitness account of how the Royal Navy struggles to operate a highly sophisticated and massively destructive nuclear armoury onboard ageing boats with insufficient experienced personnel struggling with a chronic shortage of spare parts.

Should McNeilly's account be taken at face value? When it appeared, on 17 May 2015, the Russell Foundation immediately consulted Commander Robert Green, Royal Navy (Ret'd). He agreed that the report should, indeed, be taken at face value. Commander Green went on to say about McNeilly:

> 'His revelations of serious equipment defects and lax security in Britain's sole deployed SSBN [ship submersible ballistic nuclear], and alleged failure to rectify them when reported through the chain of command, provide a timely example of how the demands of operating such a sophisticated, hazardous weapons system lead to overstretch, cutting corners, and cover-ups. The conclusion I draw is that the RN is already out of its depth trying to operate the

existing Trident system, starved of resources and trying to get by on the cheap. Lord help them if they have to take on whatever replacement the Yanks deign to let the Brits have – if they do.'

We asked Commander Green to expand his informed observations, which we publish in this issue. We also publish John Ainslie's extensive survey for Scottish CND of the safety of British submarines, which reveals a long history of fires, groundings, collisions and concealment. Written in response to McNeilly's revelations, 'Substandard' provides telling context for assessing the mega-hazard posed by UK Trident boats. That the UK government wishes to prolong the risk for another 50 years, at a cost of some £100 billion pounds, suggest that, like the Royal Navy, it may also be out of its depth in managing such complex tasks.

Brendan O'Hara, the perspicacious new MP for Argyll and Bute, which includes Faslane, has signalled that Trident nuclear weapons and associated activities will be subject to much more searching scrutiny going forward. We publish some of his probing questions, together with the brief parliamentary debate initiated at Westminster by Alex Salmond of the Scottish National Party in response to AB McNeilly's report. Jeremy Corbyn MP, frontrunner for leadership of the Labour Party, reminded the House of Commons of the humanitarian consequences of nuclear weapons; a timely remark in the year that Hiroshima and Nagasaki commemorate the 70th anniversary of the atomic destruction of their cities and populations in August 1945.

Tony Simpson

* * *

Michael Barratt Brown was a pillar of *The Spokesman*. When the magazine first appeared, in March 1970, he contributed an article, 'The British Economy: What Future?' which concluded: 'A programme of socialist advance will be put forward in a later article'. Michael was as good as his word; hundreds of contributions followed, not all of them programmatic. Over the decades, his regular reviews ranged across diverse topics such as the quality of milk and the artist and designer, Edward Burne-Jones, whose work held a particular attraction for Michael. No more will his essays arrive to brighten our mornings. We bid farewell to our comrade and friend with a range of contributions.

Out of its depth

*Commander Robert Green Royal Navy (Ret'd)**

A former operator of British nuclear weapons considers the implications of AB McNeilly's revelations for the Royal Navy and the United Kingdom's Trident submarines.

On 17 May 2015, the Scottish *Sunday Herald* revealed that a whistleblower, 25-year-old Able Seaman William McNeilly, had released online an 18-page report containing serious allegations surrounding the safety and security of the British Trident ballistic missile-equipped submarine force, based at Faslane in Scotland.

McNeilly listed security breaches, fires, leaks, floods, failed tests, false alarms, and defective equipment. The missile compartment was used as an exercise gym; and there were doubts about whether missiles could be launched. The young submariner had been so shocked by what he saw and heard at Faslane and on a recent patrol in a Trident submarine that he raised his concerns with his superiors. When they did not take them seriously, he felt he had to risk imprisonment to warn the public.

Having released his report, he gave himself up to the naval authorities. Clearly embarrassed, they publicly dismissed his complaints as 'subjective and unsubstantiated', but announced an inquiry. This concluded that his allegations were 'factually incorrect or the result of misunderstanding'. Since then, McNeilly has blogged that he witnessed almost all of them himself or read about them in manuals on patrol, while the rest were recounted to him by experienced submariners.

The Scottish Campaign for Nuclear Disarmament has now issued a damning report, *Substandard,* which appears elsewhere in this issue of *The Spokesman. Substandard* analyses McNeilly's allegations, and places them within the context of accidents and similar problems experienced by other UK nuclear submarines. It makes disturbing reading.

A month on from McNeilly's bombshell, the Royal Navy confirmed he had been 'dishonourably discharged'. He has since revealed that all charges against him were dropped; and the real reason for his discharge was the damage he had done to the Navy's image.

This episode has indeed delivered a severe blow to the image of the British Submarine Service. Hitherto, they were seen as the Navy's most professional élite, second only to special forces. The problem is that Trident submarines have no fighting role, their mission being simply nuclear deterrence, for which they need to remain undetected. Yet on 4 February 2009, the British and French nuclear-armed and powered submarines on patrol in the Atlantic farcically collided. Furthermore, their role is to await a ridiculously unrealistic, and appallingly cruel, order to launch their nuclear-tipped missiles, with the real risk of being subsequently branded as no better than terrorists.

This helps to explain the apparent lapses in motivation and professionalism. The situation has clearly been exacerbated by shortage of spare parts, and skilled personnel – because the best of both are prioritised for the Royal Navy's attack submarine force. Underlying this is the reality that, since 1962 when the British government persuaded US President John F Kennedy to let the RN have a scaled down version of the Polaris force, the British Navy has struggled to 'keep up with the Yanks'.

My conclusion is that the Royal Navy is out of its depth operating the existing Trident system, starved of resources and trying to get by on the cheap. This dangerous situation – which the courageous actions of a patriotic young whistleblower have exposed – can only get worse if the UK Submarine Service has to take on whatever replacement the US is prepared to let the British have.

Amid severe defence budget cuts, the British Army and Royal Air Force see Trident replacement as a financially vulnerable irrelevance at a time when the security focus is on the so-called 'war on terror'. The main security threats in the 21st century include climate change, poverty, resource depletion, and financial crises, as well as terrorism. Nuclear deterrence prevents rather than assists the global co-operation required to solve them.

Trident replacement was an important issue in the referendum on Scottish independence in September 2014, because British Trident submarines can only be based in Scotland. In the recent UK elections, the anti-Trident Scottish National Party won 56 out of 59 Westminster parliamentary seats in Scotland. With British public opinion divided and a significant anti-nuclear citizen movement, the final decision on Trident replacement has been delayed until 2016.

The first anti-nuclear 'break-out' by one of the five permanent member states of the UN Security Council would be sensational. With the smallest nuclear arsenal deployed in just one system, the UK is the best candidate from among them to seize this unexpected new world role, which would overwhelmingly be welcomed by the international community.

In NATO, the UK would wield unprecedented influence – with wide support from non-nuclear-armed members – in leading the drive for a non-nuclear strategy, which must happen if NATO is to maintain its cohesion. It could cause heart-searching in the former British colonies of India and Pakistan, and open the way for a possible reassessment by the US, Russia, China and France. The Royal Navy, released from a militarily useless, politically controversial and implicitly unlawful role, which it is clearly struggling with, could refocus on what it does best: conventional deterrence, protection of maritime trade, and defence diplomacy.

**Robert Green served in the British Royal Navy from 1962 to 1982. As a bombardier-navigator, he flew in Buccaneer nuclear strike aircraft with a target in Russia, and then anti-submarine helicopters equipped with nuclear depth-bombs. On promotion to Commander in 1978, he worked in the Ministry of Defence before his final appointment as Staff Officer (Intelligence) to the Commander-in-Chief Fleet during the 1982 Falklands War. Now Co-Director with his wife, Dr Kate Dewes, of the Disarmament & Security Centre in Christchurch, New Zealand (www.disarmsecure.org), an updated ebook version of his 2010 publication* Security Without Nuclear Deterrence *is available. His 2011 Frank Kelly Lecture can be viewed at* https://www.youtube.com/watch?v=k6oOdX0mnW0, *while earlier articles appeared in* Spokesman 127 (Trident Undone).

With grateful acknowledgements to the Nuclear Age Peace Foundation, which published a slightly longer version of Commander Green's article on its highly informative website (www.wagingpeace.org).

Trident at sea

The fire next time

Able Seaman
William McNeilly

There are several versions of AB McNeilly's landmark report of his experiences as a submariner onboard one of the Royal Navy's four Trident nuclear boats. This text is based on the version that appeared in The Ecologist *magazine in May 2015, and has been edited slightly in the interests of readability. Although no longer serving in the Royal Navy, AB McNeilly provides a great service to all of us.*

My name is William McNeilly. I am an Engineering Technician Weapons Engineer Submariner for the UK's Trident II D5 Strategic Weapons System. I sent this report on the 05/05/15 to every major newspaper, freelance journalists, and whistleblower I could find. It is now the 11/05/15. I've had zero acknowledgement from anyone. The penalty for releasing this will be life in prison if I'm lucky. I'm supposed to be being watched for Top Secret Clearance yet, after sending a report containing information about how close the Nuclear Weapon System is to a disaster, I've made three flights and returned to the UK with still no acknowledgement from anyone.The worst fear for me isn't prison or being assassinated, it's the fear of sacrificing everything I have just to warn the public and yet never be heard. My leave expires today. I will be hunted down for not turning up. When I'm found I will confess and will no longer be able to keep trying to warn the public. Please release this information in whichever way you can.

I need you to publish this document

This is bigger than me. It's bigger than all of us. We are so close to a nuclear disaster it is shocking, and yet everybody is accepting the risk to the public. If we don't act now, lives could be lost for generations. This document will enlighten you to the shockingly extreme conditions that our nuclear weapons system is in right now, and has been in the past. It describes different threats and events that have happened and threats that are highly likely to happen; each one individually should raise maximum concern. I need you to publish this document or send it to someone who will; please, for the safety of the people.

This report will jump between things like food hygiene and flooded toilets, the complete lack of security, floods, and a blazing inferno in the Missile Compartment, etc. My aim is to paint an overall picture of what I've seen, and to break down the false images of a perfect system that most people envisage exists.

I gathered the information by listening to O Group meetings, reading documents, conversation, briefs, listening in to conversations, and seeing with my own eyes. O Group meetings discuss the incidents onboard and plan all boat evolutions. They are held in the Navigation Centre, which is a Top Secret Compartment.

A Swiss cheese of security holes

My Top Secret clearance is only pending. I shouldn't have been able to gain entrance to that compartment, but part of my job is Strategic Weapon System navigation, so they gave me access for training purposes. At the beginning of patrol I was kicked out of the Navigation centre when O Group were about to begin, but I found a way to stay. There's a computer down the back that I worked on. Out of sight, out of mind. I could hear everything, and no one told me to leave when I was there.

This report contains references to CB8890: the instructions for the safety and security of the Trident II D5 strategic weapon system. I'm sure all the Strategic Weapon System (SWS) personnel are scratching their heads and wondering how I'm writing this on my personal laptop and referencing a book which is contained within a safe in the Missile Control Centre (MCC). The MCC is the compartment used to control the launch of the nuclear missiles. It can only be accessed by people on the access list, and no personal electronics are allowed. I was on the access list but how could I have gotten a copy of every single chapter on to my phone? A hidden camera? No. Smuggled the book out then filmed it? No.

What I did was walk into a room where no recording devices are allowed. I sat down, took my Samsung Galaxy SII (white) out of my pocket, and recorded the entire book word for word. I held the phone still, about a foot in front of my face, and anyone who looked at the screen or used common sense would've seen I was recording. There were other Strategic Weapon System personnel in the room; in the video you can see an SWS Junior Rate (JR) about three feet in front of me talking to another SWS JR sitting right beside me. You probably think that's impossible, but I've got the evidence to prove it.

The complete lack of concern for security worries me. The fact is it would've been even easier for me to cause a nuclear catastrophe than to gather that information, and gathering that information was actually quite simple, due to the amount of ignorance. We are at war, with a new kind of enemy. The terrorists have infiltrated every nation on our planet. Our nuclear weapons are a target that's wide open to attack.

To survive we must adapt

You don't have to be Alexander the Great to see we must adapt our strategies. The Cold War is over; are we still in a situation where we must invest billions upon billions into a system that puts our citizens at risk? NO! We must adapt to the evolving world in order to survive!

Here's an example of how little people outside of the Trident Programme know about it: I was part of a squad that went through basic training, and had almost finished our Phase 2 weapons engineering training before we knew we would be joining the SWS department (Strategic Weapon System). We came into contact with a lot of instructors and there was only one person who knew anything besides the names of the *Vanguard* class submarines (Trident). The only reason he knew was because he served on them. There is a strict need-to-know policy for the *Vanguard* class submarines, regardless of military rank or political authority.

I had envisioned a system with strict security and safety. I didn't see how wrong I was until I arrived at HMS *Neptune* (Faslane) and started doing the dry phase of the submarine qualification (SMQ Dry). My class sat in a room that overlooked the submarines. We all looked at the defences and contemplated how an enemy might take one out. We thought of multiple ways that one could be taken out, but they all required military grade equipment. I still thought it was as safe as it gets; no alarms bells were ringing in my head until the first boat visit.

In the classroom we were told to take all electronic devices out of our pockets, and warned that we might be searched. We headed down to the final access gate to the Green Area; the last security checkpoint, unless you count the Quartermasters (QM). I've walked past them so many times without showing identity (ID) that I don't consider them a line of defence. I'll explain those situations later on.

Harder to enter a nightclub than Faslane Green Area

At the gate the guard barely looked at my pass, which was a paper sticker with my face on it, mounted onto a piece of cardboard. The whole group throw their passes into the security office without the security officer examining them or even showing an interest in having a look to see if their faces matched the pictures. It's harder to get into most nightclubs than it is to get into the Green Area.

There's still the pin code system to get through the gate! Oh, wait. No there's not; it's broken, and anyone standing there who has thrown their security pass in, or NOT, will get buzzed through. If you have a Green Area pass, or any old green card, you can just show it to them from about three metres away (if the boat's on the first berths; if not, one metre) then get buzzed through!! That's the toughest part of the security!!

There are some security guards who will expect you to put the pass to the window so they can inspect it. However, the vast majority of them don't. We approached the Quartermaster's box to get our security brief then headed down the boat. No search at all. It wasn't because we're Royal Navy personnel; it was because that's the standard procedure.

Hundreds of contractors go down the boat when it's alongside. Their equipment isn't searched and they are not patted down. All it takes is someone to bring a bomb onboard to commit the worst terrorist attack the UK and the world has ever seen.

A perfect example of how poor the security is, was when I first got my Green Area pass and was assigned to a boat which was in the ship lift. It was a rainy dark winter's morning. The bus took us down to the gate and about 10 people were about to gain access to the Green Area at once. We all threw our Green Area North (GAN) passes into a pile in the box; without showing any ID. Then we all got buzzed through. Anyone without an ID or a GAN pass could've easily gotten through in the group.

This was not a one-time occurrence; it happens every morning. Sometimes, when it was windy and raining, I kept my face looking in the complete opposite direction from the guard so they couldn't see my face; I was still buzzed through. Anyone can catch that bus from the Yellow Area (normal base area/area before Red Area).

IDs are checked on the way into the Red Area (area before Green Area) on the bus; by a guy who just walks up and down barely looking. I've gotten through a few times by just showing my pale white room key; looks nothing like a Green Area pass. Also, if you just walk into the Red Area from the Yellow area, most of the time they will look at your ID from about 4 metres away then tell you to go on through the road part (especially if it's raining).

At a base security brief we were told that thousands of Royal Navy IDs go missing every year. A terrorist could use them, or create counterfeits with them, and easily gain access to the submarine. Considering most of the guards barely look at them from a few metres away (couple of feet if they're the rare ones), the fakes wouldn't have to be too perfect. I've shown a room card or nothing at least once at every checkpoint.

Feet away from the nuclear missiles without an ID check

I had entered the Green Area in a cluster of about 10 people and then I found out I needed to go to the boat that was in the ship lift. I was in a group of six personnel that was going to do BSS on the boat, which is basically walking around every compartment onboard learning about the boat. We went through the ship lift, past the Quartermaster, down to MC (Missile Compartment) 2 deck, set our bags just feet away from the missiles, and no one stopped us. Keep in mind, this was our first time on the boat. No one in the crew knew who we were, but they still didn't stop us. I did the same thing every morning, for the next four days.

I went through into the Red Area without my ID being checked for facial resemblance, through into the Green Area in a cluster of people, then walked straight down to Missile Compartment 2 deck and sat my unchecked bag beside the missile. I did it for almost a week before a Quartermaster stopped me to see if I was on the access list.

He told me that I should be handing my ID into the Quartermaster's box so they know who's on board if there's an emergency. One Quartermaster for one night had done his job, all the whole time I was there. Accessing a boat which isn't in the ship lift is just as easy. People very rarely get stopped by the Quartermasters unless they're in groups or look like they're lost.

You can carry anything through the security checkpoints without it being checked. When I helped with storing ship, I brought things of all shapes through and none of it was checked. Before sailing I brought my own stuff onboard in a huge grip bag; it wasn't checked. There were 31 BSQ's (submarine basic qualification) + ships staff + civilians = over 180 people bringing huge unchecked bags onboard.

If you've been through airport security after 9/11 you'll have seen how thorough the security is nowadays. If airport security and nuclear weapon security were both compared to prisons, the airport would be Alcatraz and base security would be house arrest.

Touch what you want – none of the equipment works anyway

Jumping back to my first time down the boat (SMQ Dry), I was far from impressed with the security and I was about to be extremely disappointed in the condition of the equipment. We went to the control room; the instructor said *'don't touch anything'*. A crew member responded by saying '*it doesn't matter, none of it works anyway, you can touch what you want'*. Everyone laughed. They also complained in the Missile Control Centre (MCC) about how their equipment was *'F**KED!'*.

There were a lot of red tags on equipment in most of the compartments we went into. I highly suspected a lot of them were for defect rectification, rather than standard maintenance 'tag-outs'. Seeing the condition of the security and equipment made me more than concerned for the safety of the people. It was at that point I realised I needed to gather as much safety and security information as I could. My intentions at that point were to make the changes by reporting through the chain of command.

In submarine qualifying (SMQ) dry course I learnt that HMS *Vanguard* is in the worst of the worst condition. Countless times it tried to sail but had to come back in, forcing the other boats to do extended patrols.

In one of the lessons the instructor mentioned they found a problem with one of the nuclear reactors on one of the boats. He said all boats might need to get their reactors replaced. The instructor didn't give away much information.

I knew I had to get assigned to a boat and go on patrol as soon as possible in order to gather this information. Fast track to a Leading Engineer was the answer. If I got fast tracked I would be on the first available boat after training. I worked hard day and night, and at the end of the 10 week course I had achieved the highest test result on average out of 20-plus people on the SMQ course. At the end of SMQ dry training no one received fast track. However, the achievement went onto my record. There was just one course left, one last shot: the Trident Training Facility (TTF). At the end the course I was told I had more Strategic Weapon System knowledge than most of the supervisors onboard. It was a nice compliment, but I doubt it. I was awarded fast track to Leading Engineering Technician and received an award for best student.

Just weeks after passing out of training I had a draft for HMS *Victorious*. My workmates started calling me a terrorist robot because I remembered everything and I have a Northern Ireland accent. This reputation would have undoubtedly made it difficult for me to gather information. I needed to create distance between them, and create a new persona; I aimed for a mixture of dumbness and eagerness to learn for simple curious reasons.

Within days of being on patrol I was no longer the terrorist robot, soaking up all the information for terrorist reasons. Playing dumb came easy for me; I've been doing it, and been it, most of my life. It makes people open up and explain a lot more. If someone assumes you know something they might leave that part out of the conversation, meaning you've just lost information which might have been valuable. It also helps with getting out of certain situations. I watched a lot of *Columbo* when I was a kid.

A shambles in the galley

Stores Ship – the crew was getting ready to sail. I was assisting with storing the supplies on the boat. This day gave a good indication as to how the patrol was going to be: disorganised and a risk to health. Nobody took charge of storing ship. Most of the crew, who were supposed to be helping us, left early. There was food on the ground, food thrown in the skip/bin, with wrappers busted, people throwing food at personnel on the casing, and a lot of food still waiting to be brought onboard. We had started in the morning and it wasn't until the night that the Petty Officer (PO) came out to take charge. He ordered us to bring onboard the meat, which was laying on the floor and in the bin for a good part of the day. There was meat which had dirt on it because the wrapper had busted; it was still brought onboard for us to eat on patrol.

The firefighting equipment was brought on broad at the last minute and stowed away in a rush by BSQs (non-submarine qualified personnel); most of them didn't know where to put the gear. If the suits were stored incorrectly, it could dramatically affect the response time to an incident. I also don't like the idea of removing a lot of the firefighting equipment from the submarine whilst in harbour. Their reasoning is, it's for re-entering the submarine from the casing if there's a fire. How about having sets onboard and sets at the fire dump for re-entry, so the other PPI Gold teams have the option of getting dressed anywhere onboard or from the casing? I said that to a Petty Officer and his response was *'it's a good point, they probably don't do it for money reasons'*. Considering the billions that are poured into these submarines, I doubt and hope it's not for money reasons.

Day one of patrol

It was a dark, rainy and windy morning when I made it through the gate to the Red Area with my helmet on and looking down so the guard couldn't see my face; he never asked me to look up. I made it through the Green Area checkpoint by keeping my face away from the guards, I didn't show my ID, and I never handed any ID in. I got buzzed straight through with the others. A roll call was done for the BSQ's (Basic Submarine Qualification).

It turns out there were too many people onboard and a bunch of people weren't actually supposed to be there. A few of them got asked to leave but they still kept too many, which meant there would always be two people sharing one bed ('hot bunking') because they didn't have the space. They also set up new beds, one of which blocked a major hydraulic isolation and another two blocked the port and starboard DC switchboards – three things that we need to gain access to in an emergency. The risk was recognised after about a week, and two beds at the switchboards were taken down. There were 31 BSQs on the boat; 31 extra people to get in the way of the damage control teams in an emergency; 31 people to distract watch keepers with their task books.

Initial dive – a black comedy of errors

On the first dive there was a loud continuous banging heard by everyone. It was down the forward starboard side. The next day in the junior rates mess, I heard people complaining amongst themselves about it being ignored. After all, patrol objective no.1 is to remain undetected, except by forces allocated in direct support. They suspected it might have been the fore-planes. The fore-plane is a control surface that is used to alter the depth of the submarine. There were jokes about the fore-planes being defective throughout the entire submarine. They joked about getting them stuck in dive mode. The aft-planes on full rise would compensate, if that did happen. However, would you feel safe having a plane fly over our cities that had a problem with getting stuck in dive mode? When the boat was on index they shut off from diving and stayed on the surface for safety concerns due to the fore-planes.

We need to dive whilst on patrol to remain undetected; the safety concerns were, as always, dismissed. We are down to two boats that are available for patrol. Both of which have major defects. The question is, what does it take for us to stop the Continuous At Sea Deterrent (CASD)? CASD forces there to be one submarine at sea on patrol all the time.

A problem occurred with the main hydraulic plant. I stood at the laundry where the Mechanical Engineers (MEs) hang out, to gather information. Somehow, seawater was getting into it. The amount of actual hydraulic oil in the plant had fallen to 35%; the rest was seawater. An ET ME SM (engineering technician marine engineering submarine) called the officers' plans to deal with the situation *'stupid'*. Weeks had passed and the problem was still there. I then heard a Leading Mechanical Engineer say there was an estimated 400-500 litres of seawater in the main hydraulic rep tank. The problem was there until the end of the patrol.

Hydraulics are used to open the muzzle hatches. This defect stopped them from doing a Battle Readiness Test (BRT), which proves that the muzzle hatches could have opened whilst on patrol, and that if we needed to we could've launched.

Throughout the patrol there were constant problems with both distillers. One distiller didn't work at all; the other one would produce half of the recommended output until it would also stop working, at random times. The distillers are used to turn seawater into fresh water. You would expect one of the *'most advanced submarines on the planet'* to be able to provide fresh water.

Missile control and monitoring left unmanned

I could sometimes hear alarms on the missiles' Control and Monitoring Position (CAMP) while lying in bed. I later found out that I would've been hearing them more frequently if they didn't mute the console, just to avoid listening to the alarms. This is the position that monitors the condition of the missiles, and they muted the alarms.

One of the Watch Keepers told me and laughed about how they would deal with any issues; they would deviate from set procedures because the procedures can be *'long and winding'*. He said '*sometimes you just know that you can adjust a valve slightly and that would solve the problem. Following the procedures might take you down a long and winding path.'*

You might think that's no big deal, just an engineer using his engineering skills; if he was caught doing this kind of action on an American submarine, it would cost him his job and possibly his freedom. If you work on the Strategic Weapon System you must follow the procedures; mistakes can be catastrophic.

The Control and Monitoring Position, for obvious reasons, requires constant manning. This is another rule that isn't followed. This time it's not always the CAMP Watch Keepers' fault. During the Captain's rounds period, for example, they're forced to clean a massive compartment. This draws them away from the monitoring screen and, if the alarms have been muted ... Most of the time they sit talking in a tea area; the screens aren't visible from this position. Sometimes, the MC patrol and CAMP Watch Keeper completely disregard the rule of constant manning.

I was in the Missile Control Centre when I heard a pipe saying *'could the CAMP Watch Keeper please report to CAMP'*. They got caught out that time. There were 31 BSQs (unqualified submariners) running around and distracting people in these crucial monitoring positions, such as the nuclear reactor's Main Control Desk (MCD), the nuclear missiles Control and Monitoring Position (CAMP), the control room panel, etc. These are positions which require permanent manning and permanent attention. However, those rules aren't followed. When I was doing my BSQ I could see the lack of attention they were paying. It was only a matter of time before a mistake was made, and of course one was made.

Serious electrical fire narrowly averted

A mistake was made on the panel in the control room. A small mistake from this position can cause a disaster. The fixed firefighting system, Weapon Stowage Compartment (WSC) fog spray, was accidentally activated by the control room Panel Operator. None of the electrical isolations that are required to be made were made, creating a high risk of fire in a compartment which contains torpedoes. It sprayed seawater over everything in the compartment; torpedoes, lights, torpedo monitoring panel, everything. I was called down to help with the clean up by the Coxswain. The place was flooded. Luckily there was no fire, this time.

The Coxswain exclaimed, '*I wonder why anyone wants to work here, everything is dangerous; one little thing and we're all f**ked!*' He also expressed concern about water spraying on electrics. Someone said, '*lucky it's your last patrol then*'. We all laughed. The general consensus is there's no set person for Senior Survivor. However, the Coxswain in an emergency sub sunk situation is the expert in escape, and I've read documentation that says he is the Senior Survivor. Hearing those words come from his mouth means a lot. As far as I know the control room Panel Operator got away with it. People were saying things like '*we all make mistakes*' and '*he's completely shaken up about it*'. It confuses me how someone could make an almost disastrous mistake and get away with it that easily. Anyone who turns up late for a shift gets an MAA (Master-At-Arms) and a day's wages deducted. Almost kill everyone and it's '*aww, poor guy, he's shaken up*'.

How a Trident sub was almost lost

That wasn't the only mistake made by the control room Panel Operator during my patrol. The panel also accidentally shut down the hydraulic pumps. Momentarily, we lost all main hydraulics before the emergency pump kicked in. There may have been incidents that I didn't hear about. All it takes is for them to press one wrong button in that position to cause a disaster.

A Fire Control Supervisor saw my interest in submarine disasters, so he gave me a book that contained detailed information about them. A lot of submarines have been lost due to simple accidents. If one simple mistake is made, it can be all over. You can find some of the information online, but most of it is covered up. It's only a matter of time before one of the Trident submarines is lost.

HMS *Vanguard*

HMS *Vanguard,* a Trident submarine, makes an appearance in the book for the deep depth incident. The submarine exceeded 300 metres (safe depth is 65 metres). They underestimated the weight of the submarine and didn't have enough speed for the aft-planes to create raise. The further the submarine descended the more the weight of the submarine increased due to pressure. The rate of weight increase was greater than the rate that they were pumping out water. The submarine was extremely close to being lost.

There was another incident on HMS *Vanguard* that did not appear in this book. Possibly due to the date of the book, or the fact they tried to cover it up, but everyone who serves on the Trident submarines knows that it was HMS *Vanguard* that crashed into the French submarine. I was talking to a Chief who was on the submarine at the time. He said they told him if he told anybody about it he'd face a prison sentence. However, there was an atmosphere in the room where people felt like telling their stories of near misses, plus he knew that we knew it was HMS *Vanguard*. He said

> '*We thought, this is it, we're all going to die. I was laughing my ass off at the time. I think it must have been the nerves.*'

He went on to explain what happened. The French submarine took a massive chunk out of the front of HMS *Vanguard* and grazed down the side of the boat. The High Pressured Air (HPA) bottle groups were hanging off and banging against the pressure hull. They had to return to base port slowly, because if one of the HPA bottle groups exploded it would've created a chain reaction and sent the submarine plummeting to the bottom. Luckily, the boat made it back safely for repair. There was a massive cover-up of the incident. For the first time, the 'no personal electronic devices with a camera' rule was enforced.

The HMS *Vanguard* crash didn't appear in that submarine disaster book but there's a book it will be in. I was talking to an SWS Navigation Supervisor. He told me that HMS *Vanguard* has had so many crazy incidents that they've got their own book filled with them. He said he'd read it himself, and some of the things that happened on that boat are insane. A lot of people think the boat is cursed and try to avoid being drafted to it, but maybe it's blessed, after all. It has survived all of these incidents and yet, somehow, it's still classed as operational. It won't be long now before HMS *Vanguard* makes its way to America.

Floods and missile compartment fire

In another conversation with the SWS Navigation Supervisor, I got him to tell me about his experiences onboard. He has experienced four floods and fires onboard. He told me the worst was a flood in the DC equipment space. The whole back section was submerged in electrified water, from the 10 kW motor generators. He said they were lucky they didn't follow the normal emergency operating procedure for that incident. Due to where the flood was coming from, if they had followed the normal procedure the submarine would've been lost. He also explained in detail the time he was the attack BA for a fire back aft. When he got back aft he couldn't see a thing, yet the fire was just a small one that was out before he got there. He explained how the new safety culture has created a relaxation of heaves; we no longer have heaves/drills with a blindfolded EBS. This means most people aren't prepared for being blinded by smoke, which is what happens in a real incident on a submarine.

Fire in the missile compartment, in harbour: one morning, I was standing in the Missile Control Centre waiting for a heave/drill. The Fire Control Chief started to tell everyone about the time they had a scheduled heave for 10am and a real incident happened before the safeguard rule pipe was made (SGRIF). After the safeguard rule is enforced, the pipes for fires, etc. don't have to say *'for exercise'*; it injects realism. The pipe for the fire on Missile Compartment 4 Deck was made. Everyone was waiting on the safeguard rule so they assumed the person making the pipe had forgotten to make the safeguard pipe first. They thought it was an exercise and he laughed at how they were *'tabs out'* and no real concern. He said once they arrived there they shit themselves. 4 deck was filled with smoke. There were reports of smoke on 1 deck before it reached 2 deck and 3 deck, because the smoke was so thick it had created a smoke cloud that travelled up the sides then down through those decks. The Chief said if it had been at sea there would've been about 50 dead bodies on 3 deck because of the amount of people struggling to find an emergency breathing system (EBS) coupling in order to breath.

There were jokes about how people struggle and fight to find an EBS coupling during a heave/drill, let alone a disaster with no ability to see. Plus take into account there were 31 BSQ's on this patrol, getting in the way. Some of the BSQs would've been useful, but most of them have never experienced it before and, undoubtedly, would have gotten in the way. They would've been using EBS couplings and blocking the path of experienced personnel; this happened a lot during heaves/drills. The Chief said they used all the SFU 90 fire extinguishers that could reach the fire, and they were running out of portable extinguishers.

The Fire Control Junior Rate said *'did you not consider using N2 drench?'* The Chief replied *'we were less than minutes away from N2 drenching the compartment'.* He said they finally got the fire out after using almost every portable extinguisher onboard.

The cause of the fire – toilet rolls

The fire was caused by the ship's toilet rolls being stacked from deck to deck-head the whole way along 4 deck (right beside the missiles and firing units). They reckon it was the heat from the cables that caused the fire. Nowadays, due to environmental concerns, we don't ditch trash/rubbish at sea. In numerous compartments on the boat you'll find plastic bags filled with rubbish sitting on top, underneath and beside electrical cables and equipment that generate heat. I made my concerns clear to a Leading Hand, a Chief and a Warrant Officer. I said *'there's plastic bags filled with cardboard touching cables and equipment that generate heat'.* I reminded them of the time the toilet rolls caught fire.

The first responded by agreeing, then laughing, because I said plastic and cardboard are both great materials for starting a fire. The second agreed but said there's nothing we can do about it. The third one agreed but then explained the operational advantage of not ditching rubbish; the enemy might find our rubbish on the seabed. Well, it's not funny. I can at least try to do something about it, and I think a fire on a nuclear submarine is a bigger threat to the safety of the people than some Russian boat detecting our rubbish. It's only a matter of time before they cause a fire.

Risk of igniting missile propellant

Most people I talked to who were Leading Hand and above had experienced a fire of some kind on board. The Warrant Officer had experienced multiple fires. A few times he mentioned how the belts in the fan room are just waiting for a fire. He said we'll all shit ourselves when the submarine fills with smoke in a matter of seconds. He's probably the hardest working, most switched-on guy I've ever met, but even he can't stop the inevitability of a fire.

CB8890 (0214) – *'If the HE charge is exposed to excessive heat without burning, it may become more sensitive and could cook to (non-nuclear) detonation, releasing radioactive materials and aerosols over a wide area.'*

(0219) – The chief potential hazard associated with a live missile is the accidental ignition of the first, second or third stage rocket motor propellant. If this were to happen in the missile tube with the muzzle hatch shut and locked, the pressure hull and bulkheads of the MC would burst within a matter of seconds.

Water and electricity don't mix

In addition, the missile contains a number of subsidiary propulsive and ordnance items that could cause damage to the missile and/or release toxic gases into the Missile Compartment if initiated prematurely. In some cases, this could also result in ignition or detonation of one of the rocket motors.

There was a leak coming through the roof in the Junior Rates' mess beside electrics. I asked them *'are you going to isolate it?'* They replied *'it's happened before, we didn't make any isolations last time and there was no fire.'* Even though water was travelling through the lights, they didn't feel it was necessary. It took over eight hours to stop the leak. It then returned on a few other occasions.

In the Riders' mess there was an extension cable attached to an extension cable, with clothes touching them and sometimes wet towels were hung above them. Anyone who has passed out of nursery can spot that hazard straightaway. All personal electronic devices needed to be Portable Appliance Tested (PAT) to see if they were safe for use. This rule was never properly enforced. There was only one guy onboard who PAT-tested the equipment. Two people in my mess got caught on week seven. I personally never got mine PAT tested so I could see how sloppy their enforcement was. I never got caught the whole patrol, even though I'd sit in the junior rates' mess with an unmarked plug on display for all of them to see.

Procedures bypassed to avoid triggering faulty alarm

The missiles' Control and Monitoring Position (CAMP) isn't the only place where alarms are ignored. In the Missile Control Centre there is a CAMP Fault alarm that appears quite frequently. Due to it appearing frequently, if an alarm is sounded, most of the time people won't even bother looking at the screen because they assume it's that same fault.

I heard there was a problem with the Starboard Turbine Generator (TG). I then found a letter in the bin telling the aft Engineers: '*Starboard TG combined emergency trip stop valve. Don't follow standard operating procedures (SOPs). Open it a little instead of fully so that indications are correct on the Propulsion Service Panel (PSP).*' Basically they avoided the standard operating procedures so they didn't have any alarms on their panel.

There was a problem with one of the turbine generators. We only have two TGs onboard. We need them to generate electricity from the reactor. If we lost both TGs we would've been down to using the battery. The problem with that is there was also a problem with one of the motor generators (MGs). One MG was dysfunctional. They had problems with it before and when it went faulty, instead of getting a new one, they 'fixed it' and sent the same one back, which obviously hasn't been fixed properly. There were lots of problems with the electrical generation equipment. Losing power could result in losing the submarine.

Sewage among the missiles

A good communications system is said to be the most important thing onboard. Yet we have an old speaker system that no one understands most of the time. Take into consideration that during most emergencies they talk through a breathing mask and you have a disaster waiting to happen.

If you've never seen a missile compartment before, you probably have a picture of a glistening high-tech piece of equipment in your head. Before Captain's rounds or a VIP visit it is pretty glistening, but during most of the patrol it's far from so. Missile Compartment 4 deck turns into a gym. There are people sweating their asses off between the missiles, people rowing between a blanket of s**t because the sewage system is defective, sometimes the s**t sprays onto the forward starboard missile tubes, and there's also a lot of rubbish stored near the missile tubes. Not an image you would expect of the 'most advanced weapon system on the planet'.

There were a few incidents of people in the gym dropping weights near the nuclear weapons' firing units. I heard one person joke about how he accidentally threw a weight and it nearly hit a missile's firing unit. A person was caught using a Bluetooth speaker to play music on Missile Compartment 4 deck. The Captain found out and a warning was issued over Full Main Broadcast (FMB) that all personal electronics would be banned if anyone else was caught using Bluetooth in the Missile Compartment.

Personal electronics ban not enforced

This is a quote from CB8890 (0430) – *'With live missiles embarked, the only portable radios authorised for use in the MC / AMS 2 are Cromwell Radios and fire fighter helmets with built in communications (FFHBC).'*

Electronic equipment in the Missile Compartment other than that required for safety and security must not be operating.

Personal electronics should be banned, yet the policy isn't enforced. You can bring whatever electronic devices you want onboard: laptops, phones, pads, etc. Almost everyone onboard sleeps on a level of the Missile Compartment. They use their own personal electronics right beside the missiles.

Simple rules like no e-cigs and no shaving are also not obeyed. With the ventilation constantly circulating air around the submarine, it is possible for the hairs to be picked up and cause short circuits. In the Missile Control Centre a Power Alert Alarm kept appearing and disappearing. A possible cause is something like dust or hair creating a short.

Living with toxic fumes

I was working in the Senior Rates' mess for a week. I heard them complain about the atmosphere not being in spec. For a while everyone was sleepy and then there were times people couldn't sleep. Too high or too low O_2 or CO_2 levels can cause this. Around the time of Captain's rounds people were complaining about being sleepy all the time. It could've been because of the extra work, but a lot of people were saying it was the atmosphere. Most people were using a cleaning solution which was supposed to be banned onboard. People were also mixing the cleaning agents together to create a super-cleaning agent. Someone told me they made a cocktail of cleaning products which evaporated instantly when they added hot water. I had a headache that lasted for the whole cleaning period; it went away shortly afterwards.

We had to start cleaning again for the VIP visit. The person I was cleaning with brought a bucket which had the banned substance in it, and within a few minutes my headache was back. I told him it was banned but he just cracked on. The product had a distinct smell, and it didn't take long before it was picked up. The Medical Assistant walked past and recognised the smell. He said '*that smells like Terry's Chocolate Orange, you know that's banned*', then walked off. Yet again the rules weren't being followed.

Top Secret? Who Cares?

I was in the Navigation centre listening in to a conversation between the Navigation Supervisor and a Sub-Lieutenant. They started talking and joking about how there's a complete lack of concern for Top Secret information. They shared their experiences; the Sub-Lt. talked about someone she knew who'd just leave Top Secret information lying on his bed for anyone to see.

When the guys in the Navigation Centre got used to seeing me, they let me see Top Secret information such as the precise bathymetric navigation zone (PBNZ) and the current location. There were also times when the Navigation Supervisor and the Junior Rate would leave the compartment. One time the Supervisor was off somewhere and the JR was cleaning outside the compartment; I was only in the compartment for about five minutes with the PBNZ folder and Top Secret laptop nearby. It would've been easy for me to gather the PBNZ information. To the right buyer this information would sell for millions. However, I wasn't interested in it. Releasing that information would show the positions the submarine would travel to on patrol. Any enemy could use it to help them find the submarine. It is possible for a foreign enemy to eliminate one of our submarines and get away with it.

The information I have released in this document has been carefully selected. I would never release information that I haven't considered the implications of. An idiot may say that releasing information about how open to attack we are will invite terrorism and create an increased risk to security. The truth is the threat already exists. I can sit on my ass hoping they don't find out about how ridiculous the security is or I can let our government and people know so they can make a change. I tried to make the changes from within. I expressed my concerns too many times without any action being taken.

The reason the Navigation Supervisor let me see the information was they knew there was a chance that I would come back as a Navigation Supervisor on my next patrol. However, I don't have that level of security clearance yet.

Inside the missiles

There was another incident more disturbing. I was with seven other new strategic weapon system personnel on that patrol. None of us have developed vetting (DV) security clearance, none of us had our bags checked, all of us got to see inside the missile, and a few of us got to climb inside a nuclear missile which could have had up to 12 nuclear warheads on it.

At the end of patrol we remove the missile inverters. In order to get parts of our task-books signed off we had to witness the removal. After the removal was complete, I was asked *'do you want to have a look inside?'* I climbed the ladder, put half my body inside the missile and had a look around. They pointed out explosives and said *'when you're doing this procedure don't touch them'*. If any of us were terrorists we would've been given the perfect opportunity to send nuclear warheads crashing down on the UK. A *Vanguard* class submarine can carry up to 12 warheads on each missile and has 16 missile tubes, which means there could be up to 192 nuclear warheads on a single boat at one time. Due to nuclear agreements the number would most likely be around 48 nuclear warheads; still enough to poison our atmosphere.

CB8890 (0215) – *'If RB [re-entry body] containment is breached, several radioactive and/or toxic materials may be exposed to the atmosphere. These include plutonium, uranium, lithium compounds, tritium gas and beryllium. If mixed with water, fumes or toxic gases will be generated. When installed in a Trident II D5 missile, RBs clustered around the Third Stage Rocket Motor are at risk from a rocket motor propellant fire.'*

(0216) – *'The RB could become physically damaged due to collision or fire in peacetime and in war could be subjected to splinter attack or the effects of detonation from enemy projectiles. This type of damage could also result from a successful terrorist attack.'*

(0217) – *'An accident or enemy action may cause rupture of the RB, burning or possible detonation of the HE and release of radioactive contamination.'*

The port team removing the inverters from inside the missile had removed them at almost twice the speed of the starboard team. When the port team started to work on the starboard missiles, the starboard team called them cowboys; they laughed about how much of a rushed job they had done. I didn't get to observe how the port team removed them so fast, but I did get to observe the starboard team for two missiles. Even the starboard team wasn't following the correct procedures. Normal reader worker routine was completely ignored; the worker carried out the operation from memory instead of doing it by the book.

They also joked about how the Americans do it. They said the Americans lay on top of each other and if one hand goes out of site from the other person there will be a lot of shouting their head off. I think that's better than letting a bunch of non-security-cleared people climb inside for a nosey.

Luckily, none of us were terrorists. However, the rate at which people are getting pushed through the system because of manpower shortages is scary. Strategic weapon system is so short on manpower it's unbelievable, and people are getting pushed through at an alarming rate. There are Leading Hands doing the jobs Chiefs used to do.

There is an SWS Leading Hand who still hasn't got his developed vetting clearance and he's in the position of a Launcher Supervisor. If I ignored the threat and stuck with my job, I'd be on course to be a Leader in the strategic weapon system department within a matter of months. Thirty per cent of my entire SWS task-book got signed of inside 10 minutes without me talking over or performing any of it. This is a task-book that has a limit of 18 months, 12 months for fast track.

A guy that was in my squad completed his task-book alongside in less than three months. He is now fully qualified to Watch Keeper who may have to carry out emergency operating procedures in the missile control centre (MCC) as a Fire-Control Junior Rate even though he has never been on a patrol, and hasn't even completed BSS or BSQ, BSS being the complete basic understanding of the boat.

Personnel problems

It's just a matter of time before we're infiltrated by a psychopath or a terrorist with this amount of people getting pushed through. Some of the personalities onboard are already alarming. Probably the most worrying is the SWS Junior Rate whose hobby is killing small animals. He also expressed his interest in watching dark porn, like crush porn, which is basically women stomping kittens to death while a guy masturbates. I have no idea how that guy isn't mentally discharged.

There are other people onboard who should be raising suspicion in people's minds, especially after that guy went on a shooting rampage on the A-Boat. One guy specifically; I don't want to name any names in this report so, for now, I'll just refer to him as Pole. During the morning of Captain's rounds, Pole physically attacked someone because they started cleaning too early. It happened in front of my bunk. The poor guy was verbally assaulted by Pole, and when he asked Pole *'who do you think you're talking too?'* Pole jumped out of his bed and attacked the guy. After the fight Pole threatened to kill the guy. I witnessed Pole get aggressive with five different people; he threatened to kill two of them. I was one of the people, but he never lived up to his promises of attacking me.

Pole also snapped at a Leading Hand on the last day of patrol. Pole was promised early departure on the first boat transfer because of his situation at home, but command decided his situation at home wasn't as demanding of early release as some of the other guys in the crew. Pole took a position in the queue for food ahead of the personnel who were next on watch. An SWS Leading Hand said to him *'you're not even next on, get behind me'*. Pole snapped at him, he called him a *'F**KING C**T'*, then stood behind him like a demented pit bull ready to attack. The Leading Hand said *'what's your problem? Get to the f**king back, now!'* but Pole didn't move until one of the Junior Rates escorted him to the back. After my breakfast I went up to see what the Leading Hand's response would be. With my usual dumb curious look I asked him *'what happened in the dinner line?'* He responded, '*he tried to get in front of me, even though he's not even next on watch'*. I said *'he shouted at you, didn't he?'* His reaction was *'he's just* *on edge because he never got the boat transfer'*. That was that. If threatening everyone, attacking someone, and calling a Leader a F**KING C**T whilst looking at them like you're about to kill him doesn't make you a suspect threat to other personnel then I really don't know what does.

It appeared most people had breaking points at some point on patrol. There was one guy who presented a prime example of how someone could go from saying things like *'back to back patrols won't bother me'* and *'patrols aren't that bad'* to completely losing it over a misplaced pair of flip-flops. This guy smoked electric cigarettes, which are banned onboard. His supply ran low and that's when he became aggressive. After losing his flip-flops he went berserk, throwing things everywhere, looking for them. He shouted *'everyone on this boat is a bunch of f**king re***ds'*. He punched lockers and went on shouting and banging for over ten minutes. Not once did the control and monitoring position (CAMP) Watch Keeper come over to see what the banging in his compartment was. When someone trying to sleep said *'what's your problem?'* he responded by verbally assaulting him. He told him *'get back in your f**king bed now you ...'* If this is how people react to a lost flip-flop after a tiny adjustment to their nicotine intake, then what else are people capable of doing once the electric cigarette ban is properly enforced? If the Captain was to catch anyone with electric cigarettes he would have most likely enforced the ban.

Coffee in the nuclear missile launch keyboard

I heard the Launcher and Fire-Control Supervisors whispering to each other in the missile control centre (MCC). The Fire-Control Supervisor spilled coffee on the missiles' Data Entry Subsystem keyboard. It set off an alarm. That's all the information I gathered on that incident because they were trying to cover it up so the Weapons Engineering Officer (WEO) wouldn't find out what happened.

One of the non-submarine qualified personnel (BSQ) told me he was contemplating leaving after his first patrol. He went to see the Coxswain; the Coxswain took him into his office. The Coxswain recommended he should leave. He said *'it doesn't get any better. You might as well leave now before you waste your life here.'*

A hydraulic leak was found on a missile's valve. During a battery clean the BSQs were sent down to help out. Afterwards one of them was complaining about how his fingers went numb, but little sympathy was felt by everyone because he nearly blinded someone by throwing a rag at his face. The rag just missed his eye, but that was enough to irritate it enough for him to require medical assistance from the doctor. This incident highlights the lack of concern for safety. There was no safety brief, the appropriate personal protective equipment (PPE) wasn't worn, and a careless action was committed.

When to raise a nuclear weapons safety alert for a radiation alarm? The SWS Navigation Supervisor role during a missile radiation alarm is Command Advisor. In order to learn what I need to know as Command Advisor the Navigation Supervisor asked me to watch him during a heave/drill. One thing I picked up on was people have different ideas of when the nuclear weapon safety alert should be raised. Most people think you alert inboard straightaway; whilst some people think you don't alert inboard until the radiation alarm has been confirmed as an actual leak and not a faulty alarm.

Reactor problems could take out a submarine

On my BSQ board I was asked *'what would you do if* The Sun *or someone like that got your number, and asked about the problems with the reactors or humidity, stuff like that'*. I told them *'I would just hang up'*.

The reactor situation is something I've been trying to gather information on for over a year now, but nobody wants to talk about it. If I pushed too hard for information people would've become very suspicious; which they probably already were since they asked me that question on a BSQ board. I felt I couldn't say *'so what's the problem with the reactors then?'* I did ask them about the humidity problem, though, because I had seen how bad it was. Forward dome and 1 deck WT flat were the worst; there was water dripping from the roof onto all sorts of electrics. They told me that there's a problem with the system and the condensation levels are 15 per cent higher than they should be. I then told them *'there's a pump in AMS1 that sprays water on an electrical distribution box'*. One of them said *'that's a bit of a design flaw'*. They laughed and changed the subject.

At one point on my board I was told the best way to take down a submarine. He said *'nobody ever thinks of it, but if they targeted that it would take out the whole submarine'*. I'm not going to say how on here; that's information nobody should talk about. A lot of people have had conversations with me about how easy it would be to take down the submarine. It's disturbing to know that the people serving on these boats are aware of many ways to destroy them from within. One of the biggest threats we face is suicidal attack from within. There have been suicides onboard, and on an A-boat we had a shooter kill his own work colleagues. There were some people that I served with on that patrol who showed clear psychopathic tendencies. The odds favour destruction, if no action is taken.

Nitrogen pressure below minimum

N_2 bottle group pressures dropped below 3625 per square inch (psi), the stated required minimum. N_2 drench is used to extinguish fires in the missile tube or in a compartment:

CB8890 (0438) – *'In the event of a fire within the MC or AMS 2 that cannot be brought under control by conventional firefighting methods, the MC/AMS2 can be nitrogen drenched using the procedures contained in DC documentation.'*

(0441) – *'Nitrogen bottle group pressures must be maintained above a minimum of 3625 psi in order to safeguard the nitrogen drench facility.'*

The reaction to the N_2 drench falling below these levels was *'there's nothing we can do whilst we're the on patrol SSBN'*. The cause of the leakage was still unknown when we got back from patrol. The last time I saw the pressure it was over 100 psi below the ordered minimum. I listened in to a conversation between the port and starboard crew CAMP Watch Keepers. The starboard crew CAMP Watch Keeper said '*we had a problem with ours on* Vanguard, *and they found a massive hole on the back of one of the Missile Gas reducing stations'*.

Emergency Operating Procedure

I went to the control and monitoring position (CAMP) to learn about their Emergency Operating Procedure (EOP) actions for a missile emergency. Two CAMP Watch Keepers and a missile compartment (MC) patrol were in the tea area, near CAMP. They wanted to know what I knew about their emergency operating procedures for a missile emergency. I told them what the book says they do. They disagreed with me. They were all convinced that it doesn't say in the (EOP) book that for a missile emergency in harbour, they need to check the Tam 73 fixed radiation monitor. One of them said *'I bet it doesn't say that, we never do that'*. I said *'okay, £20 it says that'*. I could see in his eyes he started to worry that he might be wrong, but that didn't last for long and he accepted the bet. I got the book out, and proved them wrong. Maybe they don't check it but this was about what their EOP book says they do for a missile emergency. This is a book they should know word for word. Their list of actions isn't even that long. It was shocking how none of them knew what the book said.

There were two prank 999 phone calls whilst on patrol. A 999 phone call sets off an alarm in the control and the whole boat has to carry out phase 1 damage control checks.

A CB8890 exam was coming up to test people's knowledge on the safety and security of strategic weapon system (SWS). I was asked *'have you read CB8890?'* Instead of saying *'yes, I have read it twice and I have a copy of it on my phone',"* I said *'I read through it quickly once and I skipped the annexes'*. Playing dumb worked out again. One of them began to tell me how to read the book. He said *'no one fully reads the book, they just know which parts are going to be questions, and they learn them'*. One Launcher Supervisor disagreed: he said *'everyone should know that book, especially SWS'*. I agree with the Launcher Supervisor; it's a book containing information about the safety and security of strategic weapon system (SWS) and we're SWS. The exam was a total farce. They told everyone most of the answers, and any answers people didn't know they just copied from the person beside them. I was in the Missile Control Centre (MCC) and I saw the Launcher Supervisor ringing up people who had missed the exam. He asked them *'pick a number between 27 and 30'*. The number they picked was their test result.

I could've got the code for the Weapons Engineering Officer's (WEO's) trigger safe. I was standing just a couple of metres behind him when he opened the safe. I didn't pay attention to the combination but the point is they seem to forget I'm not developed vetting (DV) security cleared. After, I was asked *'what do you do if you find the Firing Unit or Jettison Panel key?'* They told me the correct procedure and what they actually do. They told me about a few occasions where this has happened. If someone gets their hands on both keys, they can jettison a missile. Jettisoning without following the proper procedure to list the boat or turning off power to the Variable Energy Eject Panel (VEEP) would send the missile out of the tube at full force and it will fall back down onto the submarine. What they actually did when they found the key was they handed it back to the person who left it lying around. The normal reporting routine was aborted.

There was an excessive amount of trouble failure reports (TFRs) being filed in strategic weapon system (SWS) department due to operator and defective equipment. An example of one of the operator errors: they allowed a trainee to carry out a procedure he had never seen done before. It was a simple procedure, all he had to do was click 'yes' when he was told to click 'yes', but he clicked 'no'.

CB8890 (0207) – *'Authorised personnel work to carefully controlled and documented working practices.'*

There's a board with a list of Strategic Weapon System (SWS) defects on it in the Missile Control Centre (MCC). It is very close to being full. When we were passing in as fully qualified submariners, they told us we couldn't drink the shot of rum with the dolphin metal/badge in the glass. Considering everything that was happening around us, it made me feel like Harry Houdini getting forced to wear an extra pair of socks in case his feet get cold during a stunt.

The other crew's Weapons Engineering Officer (WEO) came on our boat after the patrol. People mocked him behind his back; they didn't like him because he showed keenness to do his job properly. There's an attitude common to most people onboard; they show hostility to anyone who works too hard.

Are subs capable of firing missiles?

At the end of a patrol final tests are done to see if the weapons system could have performed a successful launch. These tests let us know if we really were providing the UK's strategic nuclear deterrent/continuous-at-sea-deterrence (CASD). I had reached the end of my three-month patrol. It was time to do WP 186 Missile Compensation Test. The test was carried out three times and failed three times. Basically, the test showed that the missile compensation system wouldn't have compensated for the changes in weight of the submarine during missile launches. This means the missiles would've been launched on an unstable platform, if they decided to launch.

Another test was the Battle Readiness Test (BRT), which proves that the muzzle hatches could've opened whilst on patrol; if they needed to launch, they could've launched. The BRT was cancelled due to the main hydraulic system containing mostly seawater instead of actual hydraulic oil.

Basically, they're endangering the public and spending billions upon billions of taxpayers' money for a system so broken it can't even do the tests that prove it works. Five minutes before leaving the boat for leave, I walked into the Junior Rates' toilets; the whole deck was flooded in a couple inches of brown water. I tried the Senior Rates and it was the same. This summed up the system.

A lot to lose

A back aft Mechanical Engineer (ME) told me he was going to tell his family about everything that happens onboard. I said *'like what?'* He said he was going tell them about how everything onboard is nothing like you expect it to be. Everything is broken. He was on the exact same wavelength as me. We'd both witnessed a lot, but the number of events that we didn't see is what's scary. I know most people know the Trident Programme is a disaster waiting to happen, but they never tell the public. I've heard of people getting caught selling information and people writing stuff on Facebook, but I've never heard of anybody trying to alert the public.

One of the main reasons nobody tries to talk about it is they've a lot to lose. A strategic weapon system (SWS) submariner in the Royal Navy gets paid quite a lot of money. They're handing out £50,000 bonuses to keep people in the job. It's a good career for education and work experience. There's been more than a few people go into six figure jobs afterwards.

There are a lot of stories about people getting caught talking about information. People have been caught and punished for putting information on Facebook, so there is a general feeling that you wouldn't get to say much before being silenced. You're guaranteed to lose everything, if you talk; career, money, everything you own, your freedom, possibly your life, contact with family and friends. It's a lot to lose, especially if you think there's a good chance you won't get out much information, if any, before your caught, and, of course, there's good chance any information you do get out will be covered up.

After rising from the depths, I knew I had gained enough information to eliminate the biggest threat the UK faces. I also gained the knowledge that my desire to serve the people, no matter what, wasn't some fantasy. I will sacrifice everything for the people I serve. I may be losing a great job, the money, the freedom and, possibly, my life, but it's no longer the fight for those things that drives me. I'm driven by a vision of a better world!

There's still one thing that does bring a few tears to my eyes; that's knowing this might cause my family and friends any kind of emotional pain. Knowing that my Mum cooked a little too much food for dinner tonight … I can't finish describing this. Don't feel bad! I'm here through my own choice; I could've kept my career. I could've sold the information and made millions. I had choices; it's my own choice to walk this path and do what is right for the greater good of the people. A vision of a better world!

If I die, it wasn't suicide. I'm willing to sacrifice everything, but I would never use my own hand to take my life. If I'm killed and this report is made public, there will be a high chance of a violent revolution. I'm not seeking violence.

There's still a good chance of me receiving a pardon from the Prime Minister. I only released selected information. I'm not selling the information to the paper or a foreign government. I will be handing myself in to the police and my desires to serve the people are the same as the Prime Minister's. I also believe it's in the Prime Minister's best interest to release me. Prosecuting someone for alerting the people and the government to a major threat isn't a good image for someone who serves the people.

I'm releasing this information in this way because it's the only way I can be sure it gets out. I raised my concerns about the safety and security of the weapon system through the chain of command on multiple occasions. My concern couldn't have been any clearer. Multiple times I complained to people of various ranks about being able to walk straight down to the Missile Compartment with a bag on my back which wasn't checked, with the only security being lazy security guards who don't check IDs properly.

Another example: I mentioned to a Leading Hand, a Chief and a Warrant Officer about how storing rubbish in plastic bags next to equipment and wiring that generate heat will cause a repeat of the blazing inferno on 4 deck. I raised concerns for a lot of things, and not once did someone even attempt to make a change.

I see two paths in front of me. Ignore the threat, or risk everything I have to inform the government and the people. I was listening in to a conversation in the Senior Rates' mess about the VIPs we get on board at the end of patrol. One of the chiefs brought up the time he meet the Prime Minister, David Cameron. He was treated like every other VIP who comes onboard, in the sense that he was kept in the dark. Every time a VIP comes on board the boat becomes a ghost town, all off-watch personnel are to be out of site. No one can say anything bad about Trident. The focus is always maintained on why we need Trident.

I strongly believe that the Prime Minister and most people who defend Trident had no idea about how dire the situation is. This is not the time to judge what they did when they didn't know; it's about what they do now that they do know.

NB: *Some information is quoted from the Ministry of Defence's instruction document CB8890: The instructions for the safety and security of the Trident II D5 strategic weapon system.*

Trident in question

Alex Salmond MP
Jeremy Corbyn MP
Julian Lewis MP
Brendan O'Hara MP
Penny Mordaunt MP
Michael Fallon MP

On 28 May 2015, the House of Commons debated AB McNeilly's revelations about Trident, on the initiative of the Scottish National Party. We reprint excerpts. The new Parliament is much more alert to the dangers and impositions posed by Trident nuclear weapons, as indicated by subsequent written questions to ministers, which we also reprint together with the Government's sometimes repetitious replies.

Alex Salmond (Gordon) (SNP): … I want to do three things in the debate this evening. First, I want to get some detailed answers from the Ministry of Defence. We had today in a written statement from the Secretary of State for Defence 500 words of the suffocating bland complacency that typifies so much of the MOD's reaction to serious concerns (see below). I knew the Secretary of State for Defence at university and, although I might have accused him of many things, he was neither bland nor complacent. He seems to have picked up some bad habits in his tenure as Secretary of State for Defence. We want detailed answers to detailed questions this evening.

Secondly, I want to examine the lessons from the working, or indeed malfunctioning, of the reactor prototype HMS *Vulcan* at Dounreay and what that tells us about the safety concerns at the Faslane base. The difficulties that that reactor has experienced and the MOD's reaction to them give us serious cause for concern.

Thirdly, I want to examine the inherent safety concerns about nuclear reactors, made double of course by the fact that the nuclear reactors in this case are associated with nuclear weapons and tripled by the fact that the nuclear weapons are on a submarine. That tells us that there is an inherent unsafe aspect to Trident submarines. How can that be reconciled with the new political reality in Scotland, where by my count 57 Members of Parliament out of 59 oppose the renewal of the Trident deterrent in Scotland?

I say 57 not because I am expecting an imminent by-election in Scotland, but because the hon. Member for Edinburgh

South (Ian Murray) is on the record as opposing the renewal of Trident. I congratulate him on his recent promotion to the Labour Front Bench—I should say that we have been the cause of that promotion. I hope that that promotion to the Front Bench does not mean that he has undergone some mind-melding process over the last week. I hope that he will stay faithful to the commitments made to his constituents publicly on his opposition to the renewal of Trident as a nuclear device.

I want to start with the claims made by Able Seaman McNeilly—claims that are published via WikiLeaks and also through the excellent journalism of the *Sunday Herald*. Briefly, those claims—I know that the Minister will treat them seriously and give us the detailed answers that we seek—are, first, that at the final security checkpoint in Faslane naval base, no security checks of ID cards were made, that the PIN code system was broken, and that both Navy personnel and contractors were allowed access with no verification of identity. Secondly, aboard a vessel, on the missile compartment deck, no one asked for identification or checked to see whether personnel were on the list providing them with access to that part of the submarine. Thirdly, bags coming on board the submarine were going unchecked. It would be extraordinary, Mr Speaker, if we had a greater level of security in the House of Commons than might exist at that nuclear naval base.

Fourthly, the vast majority of equipment onboard may be defective. It was alleged that HMS *Vanguard* was in the worst condition and had to be recalled to port several times, forcing other vessels to do extended patrols. Fifthly, it is alleged that a problem with one of the nuclear reactors aboard one of the SSBNs [ship submersible ballistic nuclear] had been found and an instructor had suggested that all the boats might need to get their reactors replaced. We know that the process of refuelling is already under way.

It is claimed, sixthly, that firefighting equipment has been removed from the submarine while in port; seventhly, that complaints about defective equipment and safety concerns are being ignored; eighthly, that rules on constant manning of crucial positions such as the nuclear reactor's main control desk and the nuclear missiles' control and monitoring position are being ignored; ninthly, that the correct procedures to avoid a fire in the weapons storage compartment were not being followed, but no disciplinary action followed or was pursued; tenthly, that HMS *Vanguard* was nearly lost on two separate occasions, first in a deep depth incident, where the SSBN exceeded the recommended depth, and secondly when it crashed into a French SSBN. The report alleges that the extent of the latter incident has not been fully revealed.

It is claimed, eleventhly, that there have been numerous floods and fires aboard the SSBN, fire alarms are frequently ignored and concerns over fire hazards were dismissed; twelfthly, that personal electronics equipment is frequently used in the vicinity of the missile compartments, despite being explicitly banned; and thirteenthly, that standard operating procedures and safety procedures are routinely ignored across the board. The last of the main allegations in the report is that the tests carried out at the end of a patrol had to be conducted three times because they kept failing, largely due to defective equipment.

Those are just some of the allegations—or revelations—made by Able Seaman McNeilly. Of course, we have no way of knowing whether any or all of them have substance, but I would submit to the House that in the crucial matter of safety, which is clearly what is at stake, the House and the public deserve better information and a more comprehensive explanation than the 500-word written statement issued by the Defence Secretary today. That is not just an insult to this House; it is an insult to the intelligence of the general public.

Jeremy Corbyn (Islington North) (Lab) … The right hon. Gentleman may be aware that, in the previous Parliament, the hon. Member for Moray (Angus Robertson), the former Member for Lewisham, Deptford and I attended the conference on the humanitarian effects of nuclear weapons in Vienna. Most countries that took part in that conference have put on record a detailed assessment of the effects of a nuclear explosion, brought about by an accident or an act of war and the detonation of a nuclear weapon. Is he aware of any assessment made of what the effects would be on Scotland, on Glasgow, on the north of England or on Northern Ireland of a nuclear explosion, either by accident or design, in the Clyde, and what the effect would be on the wider population? Would he support such a report being sought from the UK Government?

Alex Salmond: I am aware of reports estimating the extraordinary damage that could result from such an occurrence. What I am not aware of is whether Her Majesty's Government have ever conducted such an assessment, and whether they would be prepared to do that now and to release the findings to the general public and to this House.

The second question I want to raise is what the failings in the prototype reactor at Dounreay tell us about the functioning of the reactors on board the submarines at Faslane … On 25 March this year, the Secretary of State for Defence, in a written statement, explained the decisions that had been taken on precautions following the discovery of the breaches in the cladding around one of the fuel cells at the shore test facility at Dounreay. For Members of the House who are unfamiliar with this, let me say that that has been a matter of concern for some considerable time. The breaches in the fuel cell cladding have led to the refuelling already of one of the Trident submarines and to the potential refuelling of a second one in the near future.

The prototype reactor is designed and operated at extensive level to test whether in future there might be breaches in the reactor on board the submarines. There are two aspects that should give the House considerable cause for concern. First, we have been assured by the current Secretary of State for Defence and his predecessor that when this has happened, all appropriate authorities have been informed and kept up to date with the consequences of these microscopic breaches, but I submit that the process of consultation and information is severely inadequate.

The responsible democratic body is the Scottish Environment Protection Agency, but despite the fact that SEPA was informed by the MOD within a reasonable timescale, but not immediately, it was years before the general public and this House were informed. Hon. Members may well ask why SEPA, a democratic agency reporting to Scottish Ministers, did not immediately and timeously release information, as it would in a civil nuclear incident affecting the environment. The reason is that the MOD invoked Crown immunity regarding the control and flow of information from the test reactor. I submit—I hope the Minister will reply specifically on this point—that we are past the stage where it is acceptable that the invocation of Crown immunity can conceal from the general public, for months and perhaps even years, nuclear incidents that may have a bearing on the safe operation of nuclear reactors in the Faslane base.

Secondly, the microscopic breaches in the reactor in Dounreay have resulted in that reactor being closed down. It is to be decommissioned in the next few years. That seems to me to be sensible when radiation leaks have been identified.

The difficulty in the matter which concerns this House is that that position has led to an examination of whether there should be a prototype reactor on the new generation of nuclear submarines. The conclusion that was reached, and that was in the statement of the Secretary of State for Defence this year, was that his expert panel concluded that it was a valid decision not to prototype PWR3 because there was no practical course of action that would enable a prototype facility to be built ahead of the first successor submarine.

It is bad enough that a prototype reactor is giving a signal of potential problems in a nuclear fleet, and that the Secretary of State for Defence did not timeously inform the general public of what was going on or allow the Scottish Environment Protection Agency to do its duty. That is bad enough, but to come to the conclusion that we might move to a main gate decision on renewing this nuclear deterrent without having a functioning prototype reactor which would tell us of potential problems in the new reactor is an extraordinary situation which must be inherently unsafe, unless the Minister has some information that the new reactor will be built in such a way that it does not have the failings of just about every other nuclear reactor built in recent history.

Lastly, I spoke of the new political reality in Scotland and the 57 out of 59 Members who were clearly elected as being against any decision to renew the nuclear deterrent at extraordinary cost at a time of austerity. The attitude of 57 out of 59 Members of Parliament from Scotland to next year's main gate decision will be to oppose it. That should give the Minister substantial cause for thought.

Dr Julian Lewis (New Forest East) (Con): I welcome the right hon. Gentleman back to his place in this House. May I appeal to him not to conflate two separate issues? They are indeed separate issues. One is the very real concerns about faulty operating practices leading to potential accidents. The other is the wider issue about whether or not we should have a nuclear deterrent. It might surprise him to know that many of us who believe that we should have a nuclear deterrent are as concerned as he is about the dangers of operating faults and accidents in the systems that we have.

Alex Salmond: These Members will conclude that, unfortunately, one of the consequences of having such a nuclear deterrent is having these systems in a situation which causes inherent danger. The point that I am making is that the working of the system is inherently dangerous because a nuclear reactor and nuclear weapons on a submarine are not an easy fit. The military value of this deterrent as an independent deterrent is non-existent. Its use would break international law. It is not a weapon of security, but a sign of the insecurity of the United Kingdom, believing that fading grandeur can be protected by being one of the big five in having possession of nuclear weapons … perhaps the Minister will allow me to paraphrase one of the great parliamentarians of the past. Given the political realities in Scotland, she and the Government will be making a fatal mistake if they believe that this costly trumpery, this useless, expensive, unlawful and inherently dangerous military plaything will be tolerated any longer by those on the SNP Benches, by this party or by our country.

Brendan O'Hara (Argyll and Bute) (SNP): ... The dossier compiled by Mr McNeilly makes for very worrying reading indeed. I, too, am extremely disappointed that the Minister sought to dismiss in a 500-word statement absolutely everything contained in that extremely detailed 18-page dossier. I regard that written statement as absolutely inadequate. It goes nowhere near reassuring me or my constituents about the level of safety at Faslane or onboard the UK's nuclear submarine fleet. As my right hon. Friend said, the dossier paints a disturbing picture of a lax attitude to safety and security both onshore and onboard our submarines. There are stories of people gaining access to meetings for which they do not have sufficient security clearance, bags being taken onboard submarines unchecked, and the routine use of untested portable bluetooth electronic devices in missile compartments.

I would like to highlight, in particular, a number of specific allegations concerning both the nuclear missiles and the nuclear reactors on board the Vanguard-class submarines. On page 4 of the dossier, Mr McNeilly states that he was told by one of the instructors that they had found problems with the nuclear reactor onboard HMS *Vanguard*, suggesting that all Vanguard-class submarines might have to have their reactors replaced. Is the Minister aware of problems beyond those already known about the nuclear reactors on HMS *Vanguard*, and are they similar to the 'trouser-leg' problems that affected the Resolution-class submarines?

As my right hon. Friend said, remembering the widely reported radioactive leak at the Vulcan reactor at Dounreay, and given the fact that it took two whole years for that to be made public, how can we as MPs hold the Ministry of Defence to account, and how can the public have confidence in the Minister's investigations, when it appears that bad news is released only when it can no longer be concealed? The Ministry told us that it would cost £120 million to refuel HMS *Vanguard*. Should other submarines in the class require refuelling, can she confirm that that would cost about £500 million?

On page 11 of the dossier, Mr McNeilly alleges that the reader-worker procedure for removing the inverters from the missiles at the end of their patrol was not followed. The line-by-line reader-to-worker instructions were, according to Mr McNeilly, 'completely ignored', and the removal of the inverters from the missiles became a race between the starboard-side and port-side teams, to see who could finish the job quickest. If that is an indication of the general attitude onboard towards nuclear weapons-standard operating procedures, we should all be very concerned.

On page 14 of the dossier, Mr McNeilly alleges that while he was on patrol with HMS *Victorious* the nitrogen drench in the missile compartment fell below the minimum 3625 psi required. When he asked what could be done about it, he was told:

> 'There's nothing we can do while on patrol.'

The nitrogen drench is what is used to put out fires in the missile compartment. Therefore, according to Mr McNeilly's dossier, if a fire had broken out in the missile compartment and the nitrogen drench was not working, there would have been a real danger of one or all three of the solid-fuel rockets within any or all of the missiles onboard igniting and going off in a sympathetic detonation, with potentially disastrous consequences for everyone onboard. As I have said, those are just three of a catalogue of very serious and deeply troubling allegations, yet none of them was addressed in the Minister's 500-word statement.

Mr McNeilly's dossier points to concerns about staff training and high levels of staff turnover leading to a worrying lack of suitably qualified and experienced personnel available to the service. The Royal Navy has a proud tradition, and we recognise that people make great personal sacrifices in order to serve, for which we are very grateful, but the rate at which people are being pushed through the training pipeline due to manpower shortages leading to massive staff turnover is of grave concern. There is more than a suggestion in Mr McNeilly's dossier that there is indeed a worrying shortage of highly trained and highly skilled professionals to fill those roles.

In any workplace, security and safety are paramount, but when dealing with nuclear weapons and nuclear-powered submarines, safety and security must be absolutely sacrosanct, because, to be frank, we are all just one mistake away from a catastrophe. I look forward to the Minister addressing all these points in detail, far beyond the 500 words on offer to us currently.

The Minister for the Armed Forces (Penny Mordaunt): … I will start by addressing the concerns raised by Able Seaman McNeilly. Hon. Members will be aware that my right hon. Friend the Secretary of State tabled a written ministerial statement at the earliest opportunity to update the House on this issue. I will go on to address the concerns that hon. Members have raised about McNeilly himself and his welfare, and then the wider issues that have been raised in this debate.

Before I do so, I hope that Members will permit me briefly to pay tribute to all those who are involved in the nuclear enterprise and nuclear deterrence. Whatever our different views about the merits of the capability, I hope we can all agree that we owe the men and women in the submarine service a huge debt. Their training is extremely demanding. Their deployments, by their nature, are mentally and physically challenging. What we ask of them, and what they deliver at a personal level, is truly extraordinary. The captain of an SSBN will have spent more than 10 years of their naval career underwater. An engineer is responsible for a machine more complex than the space shuttle. An able seaman will be working on a shift system, six hours on watch, six hours off, for three months at a time. There is a particular dedication in the service among all ranks. My personal belief is that those who work in the service firmly believe that the sacrifices that they are making on our behalf are worth it—fundamental to our national security. I hope I can speak for all Members of this House as I put on record our thanks for their dedication and their service.

I welcome this opportunity to address Members' concerns. It is vital that we reassure the public. That is why I sought to brief Members with a constituency interest at the earliest opportunity last week. We owe it to our servicemen and women as well, and to the civilian staff who support them, to rebut unjustified accusations against the quality of what they do. That is not to say that running such a complex operation, uninterrupted for over 47 years, is without its challenges, but the view being perpetuated of a culture of carelessness and complacency is utterly unjust. I can assure the House that neither the operational effectiveness of our nuclear deterrent nor the safety of our submariners or members of the public has been compromised. The Ministry of Defence has a responsibility to carry out its nuclear activities worldwide in a safe and secure manner. We take this, and our commitment to protect defence personnel, the workforce, the public and the environment very seriously. When managing safety, our aim is to maximise transparency while balancing the need to maintain national security.

In my first week as Minister at the Ministry of Defence, I witnessed the Royal Navy's response to the McNeilly allegations. It did not dismiss them. Each point raised was thoroughly and methodically investigated—not just what occurred and when, but why he drew the conclusions that he did. I want to place on record my thanks to the Navy for its swift action.

I appreciate that there are those who are calling for complete transparency on all that has been found, but that is simply not possible or reasonable—certainly not on the Floor of the House—given that this goes to the very heart of protecting a national defensive deterrent capability. Other channels that I shall touch on later may allow hon. Members a deeper dive into these issues.

Alex Salmond: May I point out to the Minister that the suffocating complacency of which I spoke was that of MOD Ministers in their response to serious concerns about safety? She is a new Minister, so perhaps she can blow a breath of fresh air through the Ministry of Defence and agree to withdraw the Crown immunity certificate that stops the Scottish Environment Protection Agency reporting on nuclear incidents. Will she at least allow that Government organisation, responsible to Scottish Ministers, to do its job as far as military matters are concerned?

Penny Mordaunt: I will address all the points made by the right hon. Gentleman, but I must stress that he cannot have his cake and eat it. The allegations he has made are about the safety and security of this capability. I will answer as much as I can on the Floor of the House, but I absolutely stress the dedication of the Royal Navy in addressing those concerns—keeping the deterrent safe and ensuring the security of the capability—and any suggestion that somehow there is complacency is absolutely not correct. I am pleased that the right hon. Gentleman has acknowledged that this evening.

I can assure the House that the Navy's investigation included an analysis of the service history of the boat and of the patrol report; a review of the ongoing programme of work to improve safety and security at Her Majesty's naval bases of Clyde and Devonport; one-to-one interviews with McNeilly's chain of command, his colleagues and McNeilly himself; and consultations with the regulatory and operating authorities. McNeilly's concerns proved to be either factually incorrect or the result of misunderstanding or partial understanding. Some of his concerns drew on historical, previously known events, none of which had compromised our deterrent capability and from which, where appropriate, lessons had been learned to develop our procedures as part of a continuous improvement programme.

Only one of the allegations is yet to be fully examined—that e-cigarettes were used on the submarine. I must stress that no corroboration for that has been found. Nevertheless, the chain of command is considering what further steps should be taken to ensure that it does not happen.

On the specific comments about security at Her Majesty's naval base Clyde, McNeilly's observations focus on one limited aspect of the security jigsaw—access to one internal area. There was no reason why he should have been aware of the extensive security that is layered around the controls that he experienced. Taken as a whole, I am satisfied that the overall system for security within and around the base at the time was robust and fully effective in meeting its requirements.

We are, however, not complacent. There is an ongoing programme of work constantly to review and improve security at the base. I hope that the right hon. Gentleman will send a clear message that he supports such work. He may not see the merits of the capability, but I hope that he would support its security and, where we identify the merits of further improvements to security, agree that they should be implemented.

Perhaps the most concerning aspect of this case is that McNeilly said he raised these matters while deployed, but was ignored. For his concerns, whether justified or not, to be ignored would be wholly incompatible with the leadership, the divisional system, and the safety culture we expect from the Royal Navy. We have investigated this thoroughly, including through interviews with McNeilly's former crew mates. We have not found any evidence, formal or informal, of his raising any safety concerns, even privately with those closest to him.

I will touch on the welfare of the able seamen, which I know is a concern of the hon. Member for Argyll and Bute (Brendan O'Hara). I congratulate the hon. Gentleman both on the two speeches he has made in close succession and on his appointment as his party's defence spokesman. I updated him after Able Seaman McNeilly was arrested, and I spoke to the hon. Gentleman at the time about the welfare checks that were being done on him. He was arrested by the police in Scotland on 18 May as he landed at Edinburgh airport, because he was reported as a missing person, having failed to return to duty following a period of leave. He was released the following day and passed into the care of the Royal Navy. On 20 May, he was moved to HMS *Nelson*, a shore establishment in Portsmouth, while concerns that he had raised were investigated. He has at all times been afforded the duty of care we give to all our personnel. He has been in contact with his family, and the Royal Navy has offered additional support to them should they wish to visit him.

Brendan O'Hara: Has Mr McNeilly been charged with anything? What is his current legal status? Has he been given access to legal representation?

Penny Mordaunt: I was coming to that. Mr McNeilly remains on duty as a serving member of the Royal Navy. He is not under arrest or in custody. Any restrictions that were initially placed on him for his own welfare—

namely, his having to seek permission before leaving base—were lifted as of Tuesday. He is not under arrest, in custody or charged. Our prime concern throughout the process has been his welfare.

The right hon. Member for Gordon raised wider issues regarding the fuel element breach at Vulcan. The issue with the reactor at the naval reactor test establishment was classed by the International Atomic Energy Agency as a level zero below scale incident, with no safety significance. Workers remain safe and the local community is not at risk. There was no leak outside the reactor circuit.

The MOD has made the Scottish Environment Protection Agency aware of the issue. The key point is this: had there been any safety issues, the MOD would have been the first to inform the Scottish Government and the local community. The right hon. Gentleman will be aware that the MOD works closely with SEPA, and continues to do so, given its responsibility for regulating environmental discharges from the site. Vulcan is subject to close monitoring by SEPA, and there is a robust and public formal agreement between the MOD and SEPA to ensure that we are compliant with the Radioactive Substances Act 1993.

Alex Salmond: Will the Minister undertake to revise and review whether Crown immunity should prevent SEPA from releasing information to the public?

Penny Mordaunt: The incident to which the right hon. Gentleman refers was a non-incident—there were no safety issues. I give him the reassurance on the Floor of the House that, if there were such concerns, the Scottish Government would be informed. I am going to make progress because I am running out of time.

Figures on discharges are not secret. The information is published annually in the 'Radioactivity in Food and the Environment' publication, which is available on the SEPA website. The naval reactor test establishment at Vulcan is safe and remains a low-risk site.

On the wider issues that the right hon. Gentleman raises, the protection of the UK is the Government's first duty. We are committed to the future of defence in Scotland, the capabilities based there, and the industry that supports and generates those capabilities. As my right hon. Friend the Secretary of State for Defence has said, we must plan for a major and direct nuclear threat to this country or to our NATO allies that might emerge over the 50 years during which the next generation of submarines will be in service. We know that there are substantial nuclear arsenals, and that the number of nuclear states has increased. Russia is modernising its forces, actively commissioning a new class of eight SSBN vessels, and preparing to deploy a variety of land-based intercontinental ballistic missile classes. It is planning to reintroduce rail-based intercontinental missiles. North Korea is carrying out nuclear tests and threatening more. It is carrying out ballistic missile tests in defiance of the international community.

The Government are firmly committed to renewing continuous at-sea deterrence. That capability is as relevant today as it has ever been. It is highly regrettable when inaccurate commentary leads to public concern about the deterrent. When such comments are made, we will investigate them thoroughly. When a member of our armed forces has concerns or questions, there are appropriate channels in the chain of command through which they can be raised. That must be encouraged.

I mentioned other channels that are perhaps more appropriate for the deep dive that the right hon. Gentleman seeks. I know from my experience of serving on the House of Commons Defence Committee and our work on the security and safety of the base that that is one such channel. I thank him once again for securing the debate.

* * *

Brendan O'Hara MP followed up his speech with a series of probing questions about the safety and security of nuclear weapons which are deployed from Her Majesty's Naval Base (HMNB) Clyde, commonly known as Faslane, and stored nearby at the Royal Naval Armaments Depot (RNAD) Coulport in Scotland.

Brendan O'Hara MP *(Shadow SNP Westminster Group Leader, Defence): To ask the Secretary of State for Defence, whether 43 Commando Fleet Protection Group Royal Marines is responsible for final denial at HNMB Clyde and RNAD Coulport only when nuclear weapons are aboard nuclear submarines in transport? 3 June 2015*

Penny Mordaunt MP, *Minister of State, Ministry of Defence, Minister for the Armed Forces:* 43 Commando Fleet Protection Group Royal Marines' core task is to provide military support to undertake final denial of access to nuclear weapons in addition to supporting the multi-agency force that protects nuclear weapons convoys.

Brendan O'Hara MP: *To ask the Secretary of State for Defence, what* *fires have taken place on UK (a) nuclear-powered and (b) nuclear-armed submarines since 3 December 2014.*

Philip Dunne MP, *Minister of State, Ministry of Defence, Defence Procurement*: Due to the nature of submarine operations, meticulous records are kept of all fire safety incidents, irrespective of how minor. Since 3 December 2014 there has been one small-scale fire, that was categorised as minor. It was dealt with quickly and effectively using onboard resources.

Brendan O'Hara MP: *To ask the Secretary of State for Defence, what grounding incidents involving UK nuclear-powered submarines have taken place since 2 April 2009.*

Penny Mordaunt MP: Two such incidents have been recorded since 2 April 2009: HMS *Torbay* grounded in the Eastern Mediterranean in April 2009. HMS *Astute* grounded off the Isle of Skye in October 2010.

Brendan O'Hara MP: *To ask the Secretary of State for Defence, what collisions involving a nuclear-powered submarine and (a) another submarine, (b) another naval vessel, (c) a private vessel and (d) a merchant vessel have taken place since 2 April 2009.*

Penny Mordaunt MP
Two such incidents have been recorded since 2 April 2009: HMS Torbay grounded in the Eastern Mediterranean in April 2009. HMS Astute grounded off the Isle of Skye in October 2010. (Sic.)

Brendan O'Hara MP: *To ask the Secretary of State for Defence, how many vessels transit by RNAD Coulport en route to Finnart Oil Terminal each year; and whether those vessels are boarded by Military Police during the transit.*

Penny Mordaunt MP: Between 1 June 2014 and 31 May 2015 146 tankers transited to and from Finnart Oil Terminal. We do not comment on specific security procedures as this would, or would be likely to prejudice the capability, effectiveness or security of the Armed Forces.

Brendan O'Hara MP: *To ask the Secretary of State for Defence, what assessment the Joint Terrorist Analysis Centre, Royal Navy Fleet*

Intelligence, MI5 and Ministry of Defence Police Special Branch have made of threats to security at HMNB Clyde and RNAD Coulport; what those threats are in order of the likelihood of them occurring; and what steps are being taken to address those threats.

Brendan O'Hara MP: *To ask the Secretary of State for Defence, what defences are in place to protect HMNB Clyde and RNAD Coulport from possible terrorist attack from the air.*

Brendan O'Hara MP: *To ask the Secretary of State for Defence, what defences exist against a momentum vessel attack on the Explosive Handling Jetty at RNAD Coulport.*

Penny Mordaunt MP: The Ministry of Defence works closely with wider national police, security and intelligence organisations, including the Security Service, to ensure it maintains effective protection for all defence assets against a range of threats. We do not comment on detailed intelligence or security measures to protect defence nuclear assets.

* * *

House of Commons

Written Statement made by Secretary of State for Defence, Michael Fallon, on 28 May 2015.

Safety at HM Naval Base Clyde

I am informing the House at the earliest opportunity on the investigation of each of the claims made by Able Seaman William McNeilly about the operational effectiveness, safety and security of our nuclear deterrent.

Having now completed our investigation, and having consulted with the appropriate regulatory and operating authorities, I can assure the House that neither the operational effectiveness of our Continuous at Sea Deterrent nor the safety of our submariners or members of the public have been compromised.

The Naval Service operates its submarine fleet under the most stringent safety regime, which is subject to independent scrutiny. The Naval Service does not put a submarine to sea unless it is safe to do so, and there are appropriate procedures in place to deal with any issues that may arise during its deployment. There are robust regulatory mechanisms, both within the Ministry of Defence (MOD) but independent of the Royal Navy and, externally with the Office of Nuclear Regulation, to ensure this. The MOD is also held to wider account by Parliament.

Able Seaman McNeilly published his comments following his first submarine deployment. He was under training, and his access and exposure to activities and material on board were appropriate to his security clearance. We have found no evidence that he raised any concerns with colleagues on board or with the Chain of Command: had he done so, the more senior and experienced submariners would have been able to explain how the boat operated and why McNeilly's concerns were unfounded. A number of the issues he raised did not occur during his patrol.

Most of McNeilly's concerns proved to be either factually incorrect or the result of mis- or partial understanding; some drew on historic, previously known, events none of which had compromised our deterrent capability and, where appropriate, from which lessons had been learned to develop our procedures as part of a continuous improvement programme. Only one of the allegations remains to be fully examined – the allegation that e-cigarettes were being used within the submarine. No independent corroboration of this has been found but even if it were true, there is clear evidence that their use did not put the safety of the boat at risk.

Able Seaman McNeilly was arrested having not reported for duty after a period of leave. He was released the next day, but confined to a specified location in Portsmouth while interviews were conducted. He is being afforded the duty of care that we give all our personnel, is in contact with his family, and is still in the employ of the Royal Navy.

I

When Newton saw an apple fall, he found
  In that slight startle from his contemplation –
'T is said (for I'll not answer above ground
  For any sage's creed or calculation) –
A mode of proving that the Earth turned round
  In a most natural whirl, called 'gravitation;'
And this is the sole mortal who could grapple,
Since Adam – with a fall – or with an apple.

II

Man fell with apples, and with apples rose,
  If this be true; for we must deem the mode
In which Sir Isaac Newton could disclose
  Through the then unpaved stars the turnpike road,
A thing to counterbalance human woes:
  For ever since immortal man hath glowed
With all kinds of mechanics, and full soon
Steam-engines will conduct him to the moon.

III

And wherefore this exordium? – Why, just now,
  In taking up this paltry sheet of paper,
My bosom underwent a glorious glow,
  And my internal spirit cut a caper:
And though so much inferior, as I know,
  To those who, by the dint of glass and vapour,
Discover stars, and sail in the wind's eye,
I wish to do as much by Poesy.

Lord Byron, Don Juan, 10th Canto

Substandard

The Trident whistleblower and the safety of British submarines

John Ainslie

On 17 May 2015 the *Sunday Herald* published serious allegations of safety and security weaknesses on Britain's Trident submarines.[1] The article was based on an 18-page report from Able Seaman William McNeilly.[2] McNeilly was training to be a missile technician on a Trident submarine and had been onboard HMS *Victorious* throughout its patrol from January to April 2015. He said 'the Trident programme is a disaster waiting to happen'.[3]

This report places McNeilly's allegations in the context of known safety issues with British nuclear submarines. Individual incidents on submarines have briefly attracted media attention – HMS *Astute* running aground on Skye, the collision between a British and French submarine, a nuclear powered submarine stranded in Gibraltar for one year. However, these episodes slip quickly from the public mind. The list of problems here is long, but not comprehensive, because one recurring feature of nuclear submarines is secrecy.

John Ainslie is Coordinator of Scottish CND. This report, first published on Scottish CND's website on 28 May 2015, shortly after AB McNeilly's own report appeared, placed McNeilly's revelations 'in the context of known safety issues with British nuclear submarines'. We are grateful for permission to reprint it in The Spokesman.

Summary

McNeilly's report brings together descriptions of what he saw onboard HMS *Victorious* and accounts that he heard of incidents and problems on other nuclear submarines. He outlines safety concerns, defects, security breaches and careless practice.

Official reports show that the Navy does not have enough Suitably Qualified and Experienced submariners to operate the Trident missile system or the reactors on nuclear submarines and that the greatest risk to the safety of the Defence Nuclear Programme comes from a shortage of personnel.

An overemphasis on operational requirements, at the expense of safety, has contributed to a number of submarine incidents in recent years.

The Trident missile was designed in a way that introduces a greater risk than other types of nuclear missile.

Between 2009 and 2012, fires on British nuclear submarines took place at a rate of around one every six weeks. Problems identified in a fire on one submarine in 2004 were repeated in a subsequent fire two-and-a-half years later.

There are 13 known collisions involving British nuclear submarines and 11 incidents when submarines have run aground. There have been three major generic defects with the reactor designs on British nuclear submarines. These were discovered in 1989, 2000 and 2012.

There is an unacceptable risk of a terrorist attack on a Trident submarine in the Faslane shiplift and there have been instances of sabotage on submarines in service with other navies.

The reactors on British submarines are much less reliable than those on American submarines. This means that there is a significant risk that a submarine could go into an uncontrolled dive.

Summary of allegations made by William McNeilly

Safety concerns

McNeilly repeats a report he heard of a fire in the Missile Compartment of a Trident submarine. Toilet rolls, stacked in the Missile Compartment, caught fire. This filled several of the decks of the compartment with smoke. The crew struggled to bring the incident under control and had difficulty using their breathing apparatus.

Able Seaman William McNeilly

Despite this earlier incident, McNeilly says that the risk of a fire in the Missile Compartment wasn't taken seriously. A major fire in the missile area can only be brought under control by flooding the compartment with nitrogen. However, he said that the nitrogen cylinders were significantly below the required pressure. Restrictions on personal electronic equipment, which could trigger an electrical fire, were not enforced. McNeilly told his superiors about rubbish near the missiles, which could have caused a fire, but no action was taken.

He was concerned about the risk of an electrical fire. He says no attempt was made to isolate electrical equipment after a leak was detected in the Riders' Mess (Riders are extra personnel on the vessel). There were serious problems with condensation in parts of the submarine. A sprinkler system was accidentally activated in the torpedo room. Some of the personal electronic equipment used by submariners had not been PAT tested.

Crew members who work on the Trident missile system should have a thorough knowledge of CB8890, the manual for Trident safety and security. However, McNeilly's exam on the manual was a sham. Some who missed the test were allocated results at random.

The status of the Trident missiles is monitored at the Control and Monitoring Panel (CAMP), but this was not always manned. An audible alarm on the panel was muted because it was going off repeatedly. A second recurring alarm in the Missile Control Compartment, due to a problem with power from one of the Turbo Generators, was also ignored.

One of the more hazardous operations conducted by missile engineers is the insertion of DC/AC inverters in the missiles before a patrol and their removal after a patrol. To do this they have to open a hatch in each missile tube and gain direct access to the missiles. McNeilly describes how the removal of inverters at the end of their patrol was rushed and they did not follow the written procedures or the practice used on US submarines.

Installing inverters on a Trident missile onboard a US submarine[4]

Other safety issues identified by McNeilly are:

- There was a list of defects on the Trident missile system on HMS *Victorious* and the list was almost full.
- One of the decks in the Missile Compartment was used as a gym and weights were thrown and dropped near missile equipment.
- Extra beds blocked access to DC switch boards and a hydraulics isolation valve.
- Use of banned substance in cleaning material, causing problems with fumes.

- There was an incident when a generator compartment was flooded on a submarine and this could have resulted in the loss of the vessel if it had been handled differently.

Defects on Trident submarines

McNeilly says that at the end of the patrol they tested the Missile Compensation System on HMS *Victorious*. This system should quickly restore the balance of the submarine after a missile is launched, to enable each subsequent missile to be fired. The test was carried out three times, and each time the test failed.

The missile hatches on the submarine are powered by the Main Hydraulic Plant. At the end of the patrol they should have tested that they would have been able to open the hatches if required. But they were unable to conduct the test because of seawater in the hydraulic system.

These two problems meant that they could not confirm that the submarine could have launched its missiles when on patrol.

McNeilly says that there was noise from the diving planes when the vessel submerged at the start of its patrol and that this was part of a wider problem with diving planes. Jammed planes can lead to the loss of the submarine in an uncontrolled dive.

There were problems with the turbo generators, which provide the main power source, and with one of the diesel generators, which are the back-up power source. The safety of the submarine would be compromised if both sources of electrical power were lost.

HMS Victorious *under repair at Faslane on 23 May 2015 (bank holiday weekend)*

In addition to these problems on HMS *Victorious*, McNeilly refers to defects on other submarines. He says that there are currently only two operational Trident submarines, probably due to refit and maintenance cycles, and that there are major defects on both the operational vessels.

He visited a Trident submarine in the shiplift and many of the items of equipment were tagged with red markers, either for maintenance or defects. When they were told not to touch anything in the submarine's control room, one of the crew responded 'nothing works, you can touch what you like'. Crew members manning the Missile Control Centre said their equipment was all 'f***ed'.

Security breaches

McNeilly revealed two major breaches of security on HMS *Victorious*. Despite not having DV security clearance, he was given access to Top Secret information showing where the submarine was carrying out its patrol. He also says he could have worked out the key to the Weapons Engineering Officer's safe when he watched him enter the combination. This would have given the junior crew member unauthorised access to the trigger which launches Trident missiles. In addition, McNeilly was told of an officer who frequently left Top Secret documents lying on his bed.

He says there was a lack of adequate security controlling access to Trident submarines. He suggested that it was easier to get into most nightclubs than Faslane.

He gave the following examples of lapsed security:

- The QM sentry (in sentry box at gang plank) was not an effective security check, as he routinely lets people pass unchecked.
- Pass checks and gate checks were not thorough. People are able to pass without showing face, especially when it is raining. It is possible for extra people to get in as part of a group. There are lots of missing Navy ID cards circulating.
- Electronic gate access with PIN not working.
- No checks on bags being taken onto submarine by sailors or civilians. He was able to leave his bags next to the missiles on his first visit to a submarine.

Careless practice

McNeilly described how at times, such as the loading of stores before patrol, the submarine was chaotic. At the end of the patrol both the junior ranks' and the senior ranks' toilets were flooded and he notes that this was an apt summary of the state of affairs on this deadly nuclear-armed vessel.

Staffing problems

In his report William McNeilly said 'the rate at which people are getting pushed through the system because of manpower shortages is scary. SWS [Strategic Weapon System] is so short on manpower it's unbelievable and people are getting pushed through at an alarming rate.'

Official documents show that the MOD does not have an adequate number of people operating the Trident missile system. In each of the last three years the MOD's annual reports have identified the Trident Strategic Weapon System as an area where there is a shortage of personnel, known as a 'pinch point'.[5]

Year	Pinch Point Group	Liability	Shortfall	Difference
2012	Strategic Weapon System Control & Monitoring Panel Rank: Leading Hand	30	-	30%
2013	Strategic Weapon System Rank: Senior Rate	120	20	15%
2014	Strategic Weapon System Control & Monitoring Panel Rank: OR4-OR8 [Leading Hand – Chief Petty Officer]	220	55	25%

These reports also show that there are not enough submariners who are qualified to carry out reactor operations. There has been a 15% shortfall in Category A2 nuclear watchkeepers in each of the last three years. The shortfall in Category B nuclear watchkeepers has been between 10% and 15% over the same period. There is also likely to be a shortage of suitably trained and experienced officers for some key posts, but this is not identified in the MOD's annual reports.

The 2014 report from the Defence Nuclear Safety Regulator (DNSR) says that the greatest risk to the safety of the Defence Nuclear Programme (DNP) is the lack of Suitably Qualified and Experienced Personnel (SQEP) – 'The difficulties in maintaining a sustainable community of suitable nuclear competent staff has been, and is again, raised by DNSR as the principal risk to maintaining safety in the DNP'.[7] The regulator has made similar remarks each year since 2006 and has identified the attractions of alternative employment in the civil sector as a contributing factor. The 2008/09 safety report for the Clyde Naval Base said that human factors were the principal root cause of 70% of nuclear safety events.[8]

Lack of training and experience has been a factor in a number of nuclear

Missile Control and Monitoring Panel (CAMP) Watch on a US Trident submarine[6]

submarine incidents. One reason that HMS *Astute* ran aground on Skye in October 2010 was that the Officer Of the Watch did not have suitable experience for carrying out a boat transfer in the dark in an unfamiliar area.[9] The Board of Inquiry report into the grounding also points out that lack of a chart on the bridge combined with ineffective supervision, 'effectively resulted in the control of navigation resting with a Leading Seaman'.[10]

Manning issues contributed to HMS *Triumph* running aground off the West coast of Scotland in November 2000. The Board of Inquiry found 'evidence that the supervision of very inexperienced Control Room Watchkeepers was, at times, inadequate'.[11] 30% of the crew were new to the vessel. Two officers implicated in the incident, the Officer Of the Watch and the Second Officer of the Watch had only joined the submarine shortly before it sailed.

A lack of relevant training contributed to HMS *Tireless* colliding with an iceberg in 2003. While specific training was mandatory if a submarine was due to be deployed under pack ice, there was no similar requirement for deployment in the Marginal Ice Zone.[12] Fleet Headquarters had not given adequate consideration to the hazards from icebergs in this zone and no under ice training was arranged for *Tireless* before its deployment. The submarine manual, SMP 27, gave the false impression that the submarine's passive sonar can reliably detect icebergs. The crew on *Tireless* assumed that this was correct, but their sonar system gave no advance warning of the iceberg which they hit.

The lack of trained personnel was also a factor leading to the very poor standards of radioactive waste management at Faslane which were identified in a 2009 report, following a succession of coolant leaks from submarine reactors. The Scottish Environmental Protection Agency was so concerned about poor practice at the site that a spokesman told *Channel 4 News* that if Faslane had been a civil nuclear facility they would have shut it down.

Balance between operational and safety considerations

McNeilly asserts that HMS *Victorious* was sent on patrol when it was not in a suitable condition. In a series of incidents it has been apparent that safety has been compromised because of an overemphasis on operational requirements. In December 1987, a Polaris missile was subject to 'adverse shock' during a handling accident while it was being loaded onto HMS *Repulse* at Coulport. The Board of Inquiry found that 'excessive pressure was put on staff' to undertake the operation due to 'an urgent need to exchange the missile'.[13]

There have been a number of serious accidents during Perisher courses, training submarine commanders, where the focus was on providing realistic operational training rather than on safety. In 1990 HMS *Trenchant* sank the Fishing Vessel *Antares* with the loss of four lives. Officers on the submarine were concentrating on a warship which was exercising with them and failed to take due account of the fishing boat nearby. *Trenchant* incorrectly reported that, although a net had snagged, the fishing vessel was safe. The submarine continued with its exercise and search and rescue alert was only made eight hours later.[14] In 2002, HMS *Trafalgar* ran aground near Fladda-Chuain. Navigation aids were deliberately concealed from the trainee commanders and no back-up navigation system was in place.

Fishing Vessel Antares *which was sunk, with the loss of four lives, by HMS* Trenchant *in 1990*

In May 2003, HMS *Tireless* collided with an iceberg. The Board of Inquiry concluded, 'The focus of RN submarine environmental effort is in tactical exploitation and there was insufficient focus (HS and on board) on the hazards to submarine safety presented by icebergs'.[15]

On 28 April 2004, 11 crew members refused to go to sea on HMS *Trafalgar* because they regarded the vessel as unsafe. They included 3 out of 4 safety specialists. One of those involved said there were 250 defects on the vessel.

In the early 1990s, the MOD continued to send nuclear-armed Polaris submarines on patrol despite a generic reactor fault which resulted in all other British nuclear-powered submarines being kept in port. The MOD did not fully understand the underlying problem with the reactors until two years after the issue was first noticed. Alan Clark was a Junior Defence Minister at the time. His published diaries show that safety advice was ignored, perhaps by the Defence Minister Tom King (TK) or, more likely, by Mrs Thatcher. Clark's diary entry for 31 Jan 1990 says:

> '... news is about to break concerning the trouser-leg fractures in Warspite's cooling system. This could affect every nuclear-powered submarine. The whatever-it-is Authority have already given their advice that we should 'cease to operate' them until the condition is 'rectified'.
>
> '... TK, quite rightly in my view, is continuing to keep the newer ones on station (although whether this is really his decision or was forced on him by the Lady I simply don't know). I suspect the latter because when, sadistically, I rattled him at a meeting, 'If – if there is an accident, it's not just you who resigns; the Government falls', he didn't blench[16].'

Faslane Peace Camp monitored submarine movements over the next two years. HMS *Resolution* carried out two very long patrols of around 108 days in 1990 and 1991. After the second of these patrols the submarine spent only 6 days at Faslane and then went back on patrol. There was no opportunity to repair defects identified during the patrol or to test the vessel and crew before deployment.

Overemphasis on operational factors, rather than safety, was also a factor in some of the other incidents reported below.

Missile safety

On most types of missile the nuclear warheads are placed on top of the rocket motors. Trident is different. In order to produce a long-range missile which is short enough to fit on a submarine, the nuclear warheads are placed around the third-stage rocket motor. This significantly increases the risk of a catastrophic accident. This weakness was identified by Sidney Drell in a US Government review of nuclear weapons safety published in 1990.

McNeilly quotes paragraphs from the Trident safety manual which confirm this problem. The document refers to Re-entry Bodies (RBs) which is an alternative name for the nuclear warheads.

> 'When installed in a Trident II D5 missile, RBs clustered around the Third Stage Rocket Motor are at risk from a rocket motor propellant fire'.
>
> 'An accident or enemy action may cause rupture of the RB, burning or possible detonation of the HE [High Explosive] and release of radioactive contamination.'

The manual also says that a fire can lead to the detonation of explosives in the nuclear warheads:

> 'If the HE [High Explosive] charge is exposed to excessive heat without burning, it may become more sensitive and could cook to (non-nuclear) detonation, releasing radioactive materials and aerosols over a wide area'

The risk assessments for the Faslane shiplift assume that the detonation of one missile will result in the explosion of all the missiles onboard a submarine and the dispersal of plutonium from all of the nuclear warheads.

The shiplift assessments do not take account of the possibility that a missile explosion could result in the dispersal of radioactive material from the submarine's reactor. However, the manual quoted by McNeilly suggests that this might occur. It says:

> 'The chief potential hazard associated with a live missile is the accidental ignition of the first, second or third stage rocket motor propellant. If this were to happen in the missile tube with the muzzle hatch shut and locked, the pressure hull and bulkheads of the MC [Missile Compartment] would burst within a matter of seconds'.

If the bulkhead was breached then there is a significant risk that the blast wave or fragments could damage the reactor and possibly trigger the release of radioactive material from the reactor.

Fires on submarines

McNeilly expresses concern that there was inadequate attention on HMS *Victorious* to the potential for water to cause an electrical fire.

In October 2004 there was a fatal fire onboard HMCS *Chicoutimi*, formerly known as HMS *Upholder*. This British built diesel-powered submarine had been handed over to the Canadian Navy and was in transit from Faslane.[17] Due to inadequate maintenance two hatches had to be kept open while the vessel sailed on the surface in rough seas. 2,000 litres of sea water, from a freak wave, flooded over the coning tower and into the

Fire damage to HMCS Chicoutimi *(formerly HMS* Upholder*)*

submarine. A short time later this sparked an electrical fire. Within seconds there was very thick black smoke. One officer collapsed and subsequently died. Nine crew members needed treatment for smoke inhalation.

Two-and-a-half years later, in March 2007, there was an explosion and fire on HMS *Tireless* under the Arctic icepack North of Alaska. Two sailors died. The Navy had failed to learn from the fire on *Chicoutimi* and the Damage Control and Fire Fighting (DC&FF) issues were seen a second time. The Board of Inquiry into the fire on *Tireless* said 'Many of the DC&FF lessons identified in the HMCS *Chicoutimi* incident in 2004 have been repeated in this incident'.[18] In both cases the control of fire fighting was difficult because large parts of the submarine were affected by smoke. An identical nozzle detached from a hose in both fires. There were issues with Emergency Breathing Apparatus in both cases. In addition, sailors on *Tireless* had not received adequate training to cope with the situation that they faced, in particular attempting to fight a fire from a ladder.

The fire on *Tireless* was caused by the explosion of a Self Contained Oxygen Generator (SCOG). There was a major failure in the assessment of the danger posed by these devices. The Board of Inquiry report said 'There are many systematic failings that contributed to the TIRL explosion which can be collectively viewed as inadequate risk management of the hazards that SCOGs present'.[19]

At the time of the explosion and fire, *Tireless* was operating under the ice pack, North of Alaska. It took three quarters of an hour for the nuclear-powered submarine to find a gap in the ice and surface. During this period the crew were unable to access the area of the fire because a door had buckled. The incident would have been far worse, had it not been for the actions of a surviving member of the crew in the affected area. Despite serious injuries, he was able to dampen down the flames. The Board of Inquiry report said 'The small fire caused by the explosion could easily have taken hold and a major conflagration ensued, with very serious consequences, if xxxxxx had not had the stamina and presence of mind to use all available means to extinguish them'.[20]

McNeilly said that the quality of the speaker system, used to issue safety instructions, was poor. The Board of Inquiry into the fire on *Tireless* noted that this system was inaudible due to noise from the incident. The Cromwell radios onboard were useless. Instructions were passed by word of mouth along the stricken submarine.

McNeilly questioned the rigour of fire drills, pointing out the submariners should be blindfolded to simulate the effects of a fire. The reports from *Chiticoumi*, *Tireless* and other fires show that visibility can be down to a few inches when smoke fills the compartments of a submarine. The report into the *Tireless* fire also found that emergency lighting was inadequate.

Access to Emergency Breathing Apparatus has been a recurring problem in fires on submarines. In April 1992 there was a fire on board HMS *Turbulent* at Devonport. Maintenance work was being carried out on one of the two electrical switchboards when there was a short circuit and a bang followed by a fire. The switchroom is adjacent to the reactor compartment and separated from it by a bulkhead. The Mechanical Engineering Articifer (MEA) of the Watch was not wearing a face mask when he was required to carry out an essential safety task. Petty Officer Christian Checkley removed his face mask and handed it to the MEA. The essential safety task may have been to shut down the reactor. The reactor was producing power at the time of the fire but was quickly shut down. The consequences of the accident might have been much worse if Petty Officer Checkley had not taken this action – for which he received the Queen's Commendation for bravery. The incident was officially described as 'potentially lethal' and 23 sailors were admitted to hospital suffering from smoke inhalation.

On 19th August 1993 at Devonport, toxic diesel exhaust fumes spread through part of HMS *Torbay*. All the 32 sailors who had been on board were taken to hospital. 13 were kept in over the weekend and some were still suffering from the effects of the accident several weeks later. Commenting on the incident, Captain Richard Sharpe, editor of *Jane's Fighting Ships* said: 'We are dealing with an incredibly small hull which is machinery intensive. The smallest amount of smoke spreads with amazing rapidity'.

Lists of fires

In 2009, the government disclosed that there had been three fires which required external assistance since 1987:[21]

Date	Vessel	Location	Notes
17 February 1992	HMS Renown	Clyde	
30 April 1992	HMS Turbulent	Devonport	Explosion and fire in switchroom; 24 casualties from smoke
24 October 2003	HMS Trafalgar	Devonport	

In 2009, the government said there had been 20 medium scale fires (which were brought under control using ship's resources) since 1987.[22] These were defined as 'a localised fire such as a failure of mechanical equipment creating smoke and flame requiring use of significant onboard resources'.[23] This parliamentary answer included one fire in 1984 and is show below.

Date	Vessel	Location	Notes
10 December 1984	HMS Courageous	Alongside	
29 July 1987	HMS Sceptre	Not recorded	
26 August 1987	HMS Conqueror	Alongside	Engine room damage and burns casualties
15 November 1987	HMS Renown	Not recorded	
10 October 1988	HMS Renown	Clyde	
5 August 1989	HMS Valiant	Clyde	
22 December 1989	HMS Valiant	Alongside	
21 November 1991	HMS Trenchant	At sea	
3 November 1992	HMS Superb	At sea	
11 January 1993	HMS Tireless	At sea	
29 July 1993	HMS Revenge	At sea	
22 October 1993	HMS Tireless	Alongside	
7 April 1994	HMS Sovereign	Rosyth	
22 August 1995	HMS Sovereign	Rosyth	
16 October 1995	HMS Victorious	Alongside	
18 January 1999	HMS Talent	Devonport	
17 June 2001	HMS Sovereign	Clyde	
22 April 2002	HMS Victorious	At sea	
11 October 2006	HMS Vigilant	Clyde	
21 March 2007	HMS Tireless	At sea	Explosion and fire; Two fatalities

Known fires on British nuclear submarines prior to 1987[24]

Date	Submarine	Location	Notes
1963	Valiant	Barrow	Fire in reactor compartment while under construction
1965	Dreadnought	Rosyth	Fire in control room
September 1968	Valiant	Chatham	Two small fires

Date	Submarine	Location	Notes
August 1970	Resolution	Rosyth	Fire in control room
3 July 1972	Repulse	Rosyth	
January 1975	Repulse	Faslane	Fire from equipment overheating
July 1975	Courageous	Faslane	
1976/77	Repulse		Fire causing £200,000 damage
2 May 1976	Warspite	Liverpool	Major fire lasting 5 hours, 1 seriously injured and 4 others taken to hospital, 2 years to repair
March 1980	Revenge	Faslane	Electrical fire, jetty cable
18 September 1983	Conqueror	Devonport	
September 1985	Repulse	Rosyth	Fire on jetty heating system
1986	Splendid	Devonport	Fire in generator
26 August 1987	Conqueror	Devonport	Fire causing engine room damage and burns casualties

Between 1987 and April 2009 there were 213 small scale fires on British nuclear submarines.[25] By July 2014 this had risen to 243; of these 67 were on ballistic missile submarines (Polaris/Trident).[26] This means that between April 2009 and July 2014 there was a fire on average once every 6 weeks. Between July 2012 and November 2014 there were a further 14 small scale fires on submarines.[27]

Fire on HMS Astute, *Barrow, 18 April 2009*

Collisions and Groundings

Collisions

McNeilly reports that details of the collision between HMS *Vanguard* and *Le Triomphant* in February 2009 are a closely guarded secret. He recounts that a Chief Petty Officer, who had been on *Vanguard* at the time of the crash, told him 'We thought, this is it, we're all going to die' and explained that the French submarine had taken a chunk out of the front of *Vanguard*, grazed down the side of the boat and dislodged High Pressure Air bottle groups.

The Nuclear Information Service submitted a Freedom of Information request for reports of the collision. A heavily redacted version of several documents was released. This gives no indication of the circumstances or effect of the collision. It confirms that both submarines were on patrol at the time of the incident.

HMS Vanguard *entering Faslane shiplift at night after collision with* Le Triomphant

Excessive secrecy surrounding nuclear submarine collisions is not new. In 2013 a submariner described for the first time a collision between HMS *Warspite* and a Russian submarine in October 1968. Ian Wragg said 'There was an almighty bang and the boat rolled 360 degrees over. Nobody really knows why it happened, but most people feel that the Russian boat had slowed down and we ran into the back of it. We were all given a tot of rum[28].' *Warspite* limped back to British waters and came into Lerwick with a damaged coning tower. Official reports said that the vessel had collided with an iceberg.

Groundings

There have been a number of occasions in recent years when British nuclear submarines have run aground. Common features in these incidents have been a lack of navigation skills and poor communications.

On 19 November 2000, HMS *Triumph* ran aground off the West coast of Scotland. The submarine had just completed a five month deployment when the submarine was sent out again, from Devonport, on a submarine command training course. Because the original crew had already been at sea for so long 30 % were replaced with personnel from other submarines.

The submarine was sailing towards the North Channel. They were approaching the continental shelf where the seabed rises to a depth of around 200m. A plan was prepared for the submarine to rise to 100m when they approached the shelf. The officers on duty did not use the automated Submarine Navigation and Processing System (SNAPS) properly. They also failed to make full use of the echo sounder to detect the rising seabed. As a result the submarine was 2.6 nautical miles away from its estimated position. The submarine hit the seabed at a depth of 200m and speed of 20 knots. Emergency Stations was piped. Then the main ballast tanks were blown and the submarine was brought to the surface.

The Board of Inquiry concluded that

> 'The grounding was caused by poor navigation. Contributory factors included a widespread misunderstanding of SNAPS organisation and poor chartwork[29].'

Ten days after this incident, on 29 November 2000, the Trident submarine HMS *Victorious* ran aground on Skelmorlie Bank in the Clyde estuary. As submarines depart from the Clyde a Towed Array Sonar system is attached by a long cable. HMS *Victorious* had planned to attach the Towed Array near Bute. However, when they approached Bute the wind was too strong. They decided to change course and go to the North of Cumbrae. When manoeuvring in this area a rope almost became tangled in the submarine's propulsor. So a third option was chosen. The submarine would sail to Loch Long and check the situation there. No clear plan was made for the move from Cumbrae to Loch Long. The Petty Officer who was logging the submarine's position was unaware that the submarine had increased speed to 10 knots. There was then an error in taking bearings to determine the vessel's position and the wrong course was selected. Skelmorlie bank is a sandbank in the Clyde estuary which is clearly marked with a large buoy. However, no-one on the submarine realised that they were sailing towards the wrong side of the buoy. They were 400 yards on the wrong side of it when the submarine grounded. At that point they were around 2.2km from the shore. Emergency Stations were sounded. The submarine then returned to Faslane.

The Board of Inquiry concluded:

> 'The primary cause of the grounding was a failure of standard navigational practice and a lack of awareness amongst all members of the navigation team of the increased danger to safe navigation as soon as the submarine had deviated from its pre-briefed navigation plan[30].'

On 6 November 2002, HMS *Trafalgar* ran aground on Fladda-Chuan, 6 kilometres north of Skye, while taking part in a submarine command course. Navigation aids had been deliberately concealed. This was done 'to increase the degree of navigational difficulty and hence pressure on the students'.[31] There was confusion on the submarine as to who was formally responsible for navigation. The trainee in control was using tracing paper which concealed some of the information on his chart. He did not properly take account of the significant tidal stream around Fladda-Chuain. As a result the submarine struck the bottom heavily at a speed of 14.7 knots. The Board of Inquiry concluded that HMS *Trafalgar* grounded because of human error. The Commanding Officer and the course teacher were both disciplined in a Court Martial.

The inquiry report said 'Although a safety organisation was in place [on the submarine] and had worked effectively up until then, it failed to operate when most needed.'

The report said that there was 'good reason' for depriving the trainee commander of navigation aids and did not recommend that this practice should cease. Submarine Command Courses have continued to take place using nuclear submarines off the West coast of Scotland. The training deliberately places the trainees in difficult and confusing situations. The trainees control the submarine as they conduct a series of exercises, some of which are at high speed and close to other vessels.

On 26 May 2008, HMS *Superb* grounded on an underwater mountain in the Red Sea. A late decision was made to carry out the traverse at depth, without adequate attention to navigation. The Navigation Officer had failed to make full use of the two charts of the area which were available.[32] As a result of the damage, *Superb* was scrapped a few months after the incident.

On 22 October 2010, HMS *Astute* ran aground while conducting a boat transfer near the Isle of Skye. On three occasions the Officer of the Watch ignored instructions from other crew members to change course. It was several hours before the vessel could be refloated. During the attempt to dislodge the submarine it collided with the tug which was assisting it.

The Board of Inquiry found,

> 'Preparations for, and conduct of the watch by the Officer of the Watch fell short of the standards required to maintain submarine safety. The planning for the boat transfer was insufficient to ensure safe completion and lacked appropriate command oversight'.

Communications between the bridge and the control room were poor. There were also problems communicating with MV *Omagh*, the vessel involved in the boat transfer.

Lists of collisions and groundings

In 2009 the government published a list of 13 incidents where submarines had collided with other vessels or ran aground since 1988.[33] This can be supplemented by lists compiled by Scottish CND, reports of collisions with Russian submarines and other sources.[34]

The following is a combined list of known collisions.[35]

Date	Submarine	Location	Notes
October 1968	Warspite		Collision with Russian submarine, crew told to say they had hit an iceberg
7 October 1969	Renown	Kintyre	Collision with Irish MV Moyle while surfacing at night
1969	Revenge	Clyde	Collision with cattle boat while surfacing
January 1973	Repulse/ Revenge	Faslane	Collision while leaving dry dock, Repulse hydroplanes damaged
1981	Sceptre	Barents Sea	Collision with Russian submarine, crew told to say they had hit an iceberg
10 June 1985	Resolution	Florida	Collision while at US missile range
24 December 1986	Splendid		Collision with Russian submarine, towed array lost.
2 July 1988	Courageous	North Channel	Sank yacht Dalriada
November 1990	Trenchant	North of Arran	Sank Fishing Vessel Antares
13 May 2003	Tireless		Collision with iceberg in Marginal Ice Zone at depth of 60 metres.
4 February 2009	Vanguard		Collision with Le Triomphant, both submarines on patrol
22 October 2010	Astute	Skye	Collided with tug after running aground
April 2015	Talent		Collision with iceberg, £500,000 damage to coning tower.

HMS Astute *aground on the Isle of Skye 22 October 2010*

HMS Repulse *aground on the sand at Barrow after launch*

Below is a list of occasions when British nuclear submarines are known to have run aground.[36]

Date	Submarine	Location	Notes
5 November 1967	Repulse	Barrow	Ran aground after launch
17 April 1971	Renown	Clyde	Hit sea bed in post refit trials, Captain court-martialled
13 October 1989	Spartan	Loch Linnhe	
March 1991	Valiant	North Norwegian Sea	
July 1996	Trafalgar	Off Isle of Skye	
July 1997	Trenchant	Off coast of Australia	
19 November 2000	Triumph	West coast of Scotland	Hit sea bed at depth of 200 m and speed of 20 knots
27 November 2000	Victorious	Clyde	Ran aground on Skelmorlie Bank
6 November 2002	Trafalgar	North of Skye	Hit seabed at 15 knots
26 May 2008	Superb	Red Sea	Hit underwater mountain. As a result the vessel was scrapped.
April 2009	Torbay	Eastern Mediterranean	
22 October 2010	Astute	Skye	Aground on Skye for several hours

Reactor Defects

McNeilly alludes to reactor problems which might result in reactors having to be replaced. This probably refers to the core cladding failure on the prototype submarine reactor at Dounreay in 2012. This incident was kept secret, including from the Dounreay Stakeholder Group, for over a year. The incident has a potential impact on all current British nuclear submarines. As a result of the problem, an additional nuclear refuelling has been arranged for HMS *Vanguard* and may be needed for HMS *Victorious*. This is only one of a series of major defects in the reactors on British nuclear submarines.

In 1989 cracking was discovered in the primary coolant circuit of the reactor on HMS *Warspite* during a refit. In January 1990, Scottish CND received a number of anonymous phone calls from someone working in Faslane. We were told 'there are cracks around the watery leg pipework in the primary circuits in the SGs [Steam Generators] in SSN and Polaris boats and they don't know how to fix them'. When asked how dangerous this was, the caller replied 'Let's just say we're talking about Chernobyl'. In a further call, we were told 'the emergency cooling system doesn't work at all. There was an incident at Faslane about two years ago when there was a near meltdown'. Hunter-killer submarines were initially confined to port, but Polaris submarines continued to be sent out on patrol. Checks were carried out on all nuclear submarines and several were scrapped as a result.[37]

On 12 May 2000, there was a leak of coolant from the reactor on HMS *Tireless* when the submarine was deployed in the Mediterranean. Experts in the UK wrongly advised the crew to restart the reactor, which made the problem worse. The vessel limped to Gibraltar where a year-long repair was carried out, promoting serious concerns from both the Spanish government and the authorities in Gibraltar. This was also a generic fault which could affect all submarines with PWR1 reactors. Checks and repairs were conducted over several years.

HMS Tireless *stranded in Gibraltar August 2000*

Terrorism Risk

McNeilly was particularly concerned about the risk of a terrorist attack on a Trident submarine and the lack of adequate security arrangements, including the failure to check bags that were taken onboard by sailors and civilian staff.

The MOD treat the threat of a terrorist attack differently from the hazard of nuclear accident. One example is the Faslane Shiplift. Risk assessments for the shiplift show that if a large aircraft collides with the shiplift while there is an armed Trident submarine inside, then the building will collapse and all of the missiles will detonate, scattering lethal plutonium over a wide area. However, the accident assessment says that this risk is acceptable because the probability that an aircraft will accidentally fall out of the sky and land on this particular part of the earth is very remote. Common sense would suggest that the risk of a terrorist deliberately crashing an aircraft into the shiplift is much higher. The compliance criteria for the shiplift show that there is not a huge margin between the calculated risk of an aircraft accidentally causing the facility to collapse and the required safety criteria. This would suggest that the risk of a terrorist incident causing a nuclear accident at Faslane is unacceptably high.

Faslane Shiplift

Sabotage

McNeilly raises the prospect of deliberate sabotage of a Trident submarine by someone onboard. One example, which McNeilly quotes, of the potential for extreme behaviour on a nuclear submarine was the shooting dead of one officer and wounding of a second by a sailor on HMS *Astute* in Southampton in 2011. There are several examples of submarine sabotage from other parts of the world.

In 2001 Ernesto Cimminio, a Petty Officer in the US Navy, was charged with deliberately damaging more than 100 cables on the Trident submarine, USS *Alaska.*[38] Cimminio was abusing drugs at the time and having an affair with the wife of another sailor.[39] There were also reports that valves were deliberately shut when they shouldn't have been and that reactor control valves were cut on USS *San Juan* in 1996.[40]

The destruction of the Indian submarine INS *Sindhurakshak* in 2013 has been reported as likely to have been the result of sabotage. There was an explosion and fire which led to the detonation of torpedoes stored on the vessel. All 18 crew members who were onboard were killed.[41]

Uncontrolled dive

McNeilly says that there were problems with the diving planes on HMS *Victorious*. He said 'There were jokes about the fore-planes being defective throughout the entire submarine. They joked about getting them stuck in dive mode'. The diving planes on British nuclear submarines are critical. He also refers to problems with the diesel generators. A combination of reactor, diving plane and generator problems could be catastrophic on a submarine operating close to maximum depth.

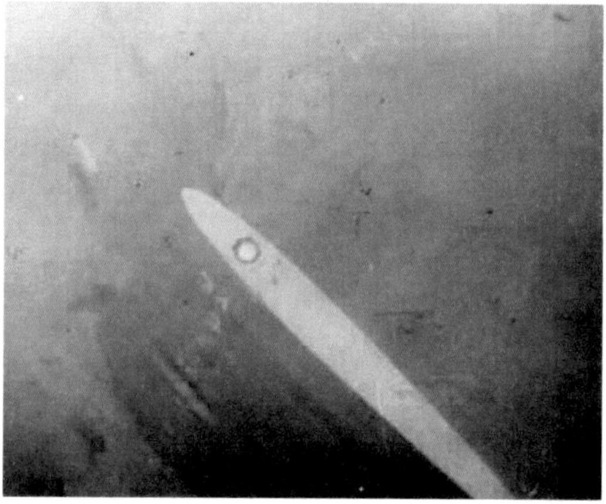

Tail fin of USS Thresher on the seabed

On 9 April 1963 the USS *Thresher* was conducting trials when it went into an uncontrolled dive. The vessel sank to a depth of 2,600 metres and was lost with all 129 crew members. A Court of Inquiry concluded that the reactor had probably shut down, resulting in a loss of propulsion, and that the ballast system had also failed, possibly as a result of freezing temperatures.[42]

Today the US Navy has design principles to prevent a recurrence of this disaster. However, current British submarines fall short of these standards. This was revealed when the MOD failed to properly redact an electronic document issued to Scottish CND under the Freedom of Information Act.[43] The report from the Defence Nuclear Safety Regulator (DNSR) said that the reactors on British submarines are much less reliable that the reactors on American submarines. The hidden text in the document said:

> 'US established practice is to deliver a high reliability of propulsion, from the main propulsion system, even under reactor fault conditions. UK practice in current class submarines is to accept a much lower reliability from the main propulsion system, and to back this up with a very low power (but high reliability) emergency propulsion system. This system will not provide

> sufficient dynamic lift, so safety is achieved by procedural controls constraining the combinations of speed and depth, backed up by ballast systems (but this may not be effective under all circumstances).'[44]

This means that, whereas safety is engineered into US submarines, on British submarines safety is based on personnel following procedures. The DNSR report revealed that, when compared with US standards, 'it is clear that the UK programme currently falls short of current relevant good practice.'[45]

A British submarine faced with a loss of power at close to maximum depth could be placed in an uncontrolled dive without the means of regaining the surface. McNeilly recounts an incident on HMS *Vanguard* when the submarine was 'extremely close to being lost'. This is almost certainly a reference to a near disaster on the submarine when operating in the Celtic Deep at close to maximum depth in July 1998. A sailor contacted the *Sunday Mail* and told them 'The boat was shuddering and shaking. We were on our knees praying. Everyone was scared out of their wits because we had never experienced anything like this'.[46]

Apparently, the reactor on HMS *Vanguard* had shut down. The crew tried to get a back-up power system going but it failed to work. By this time the submarine was descending rapidly in a deep dive. The vessel was eventually brought back under control when the reactor started up again. John Large, a nuclear engineer, said 'This is an extremely serious problem for this class of submarine. It's horrific – the worst nightmare.' The Royal Navy admitted that HMS *Vanguard* had been forced to make 'an unscheduled surface during a training exercise', but denied that there was any cause for concern.

Notes

1 http://www.robedwards.com/2015/05/trident-is-a-disaster-waiting-to-happen-says-nuclear-whistleblower.html

2 http://robedwards.typepad.com/files/the-nuclear-secrets.pdf

3 ibid

4 The photo shows two sailors on top of each other. McNeilly says the UK wasn't following this practice. http://navy.togetherweserved.com/usn/servlet/tws.webapp.WebApp?cmd=ShadowBoxProfile&type=TrainingExt&ID=92862

5 MOD annual report and accounts 2013/14 https://www.gov.uk/government/uploads/system/uploads/attachment_data/file/381341/MOD_AR14_print_version.pdf

6 https://ricochet.com/archives/sleep-when-youre-dead/

7 https://www.gov.uk/government/uploads/system/uploads/attachment_data/file/

407154/20150122-DNSR-Annual-Report-2013-14-Issue-FINAL.pdf
8 http://robedwards.typepad.com/files/shefannual-report-0809.pdf
9 Board of Inquiry into the grounding of HMS *Astute* 22 October 2010 https://www.gov.uk/government/uploads/system/uploads/attachment_data/file/27118/astute_grounding_si_report.pdf
10 ibid
11 Board of Inquiry – Grounding of HMS *Triumph* on 19 November 2000
12 Board of Inquiry into the collision of HMS *Tireless* on 13 May 2003, Faslane Flotilla, 5 June 2003.
13 RNAD Coulport Jetty Missile Loading Incident 3 December 1987 Report of the Board of Inquiry, 14 December 1987.
14 Report of the Chief Inspector of Marine Accidents into the collision between the Fishing Vessel *Antares* and HMS *Trenchant* with the loss of four lives on 22 November 1990.
15 Board of Inquiry into the collision of HMS *Tireless* on 13 May 03, Faslane Flotilla, 5 June 2003.
16 *Diaries: In power 1983-1992*, Alan Clark
17 http://www.crs-csex.forces.gc.ca/boi-ce/rp/hmcs-ncsm/rp/index-eng.aspx
18 Board of Inquiry into the circumstances surrounding the deaths of LOM(WSM) Paul David McCann D252555M and OM(WSM)2 Anthony Huntrod D261007R and the injury to xxxx of HMS *Tireless* on 20 March 2007 http://www.banthebomb.org/images/stories/pdfs/boitireless.pdf
19 ibid
20 ibid
21 Hansard 16 September 2009 Col 2223W
22 ibid
23 Hansard 2 April 2009 http://www.publications.parliament.uk/pa/cm200809/cmhansrd/cm090402/text/90402w0024.htm09040272000014
24 Cracking Under Pressure, Scottish CND & Faslane Peace Camp 1992 http://www.banthebomb.org/archives/magazine/crackin2.htm
25 Hansard 2 April 2009 http://www.publications.parliament.uk/pa/cm200809/cmhansrd/cm090402/text/90402w0024.htm09040272000014
26 Hansard 3 July 2012 http://www.theyworkforyou.com/wrans/?id=2012-07-03c.113754.h"http://www.theyworkforyou.com/wrans/?id=2012-07-03c.113754.h
27 Hansard 3 December 2014 http://www.parliament.uk/business/publications/written-questions-answers-statements/written-question/Commons/2014-11-26/216078/
28 http://www.chad.co.uk/news/local/ian-relives-nuclear-sub-crash-terror-1-6149498
29 Board of Inquiry – Grounding of HMS *Triumph* on 19 November 2000
30 Board of Inquiry into the circumstances surrounding the grounding of HMS *Victorious* on Skelmorlie Bank on 29 November 2000
31 Board of Inquiry into the Grounding of HMS *Trafalgar* on Fladda-Chuain on

6 November 2002, HMS *Montrose*, 12 November 2002
32 Board of Inquiry report into the circumstances surrounding the grounding of HMS *Superb* on 26 May 2008 https://www.gov.uk/government/uploads/system/uploads/attachment_data/file/27140/boi_hmssuperb.pdf
33 http://www.publications.parliament.uk/pa/cm200809/cmhansrd/cm090402/text/90402w0024.htm09040272000014
34 Cracking Under Pressure, Scottish CND and http://whatreallyhappened.com/WRHARTICLES/KURSK/kursk.html
35 Cracking Under Pressure, Scottish CND and Hansard 2 November 2010 http://www.publications.parliament.uk/pa/cm201011/cmhansrd/cm101102/text/101102w0001.htm
36 Cracking Under Pressure, Scottish CND and Hansard 2 November 2010 http://www.publications.parliament.uk/pa/cm201011/cmhansrd/cm101102/text/101102w0001.htm
37 Cracking Under Pressure, Scottish CND & Faslane Peace Camp 1992 http://www.banthebomb.org/archives/magazine/cracking.htm
38 http://web.kitsapsun.com/archive/2001/06-02/0006_uss_alaska_damage__navy_drops_sab.html
39 http://www.seattleweekly.com/2001-02-28/news/slicing-the-sub/
40 https://news.google.com/newspapers?nid=1915&dat=19971008&id=MZdGAAAAIBAJ&sjid=evgMAAAAIBAJ&pg=2083,1506820&hl=en
41 http://www.sunday-guardian.com/news/submarine-fire-may-have-been-sabotage
42 Navy News 26 October 1993 http://www.strategypage.com/military forums/462-663.aspx
43 http://www.banthebomb.org/images/stories/pdfs/substandardsubmarines.pdf
44 Successor SSBN – safety regulator's advice on the selection of the propulsion plant in support of the future deterrent review note. DNSR, 4 November 2009.
45 ibid
46 http://www.banthebomb.org/archives/magazine/nfs989o.htm

* * *

God gave Noah the rainbow sign,
No more water, the fire next time.

Uncommon Cause

General Lee Butler USAF (Ret'd)

Interviewed by Robert Kazel

General Butler first appeared in The Spokesman *(no.63) in 1997, in the company of General Andrew J Goodpaster, engaged in 'Eliminating Nuclear Weapons'. The former chief of US Nuclear Forces holds fast to this aim, as he explains to Robert Kazel. A longer version of their exchange is available online at the website of the Nuclear Age Peace Foundation, to which we make grateful acknowledgements. General Butler's* Uncommon Cause: A Life at Odds with Convention, *will be published this year.*

General Lee Butler, pictured here in 1983, was commander of the 320th Bombardment Wing, Mather Air Force Base, California. The base was home to long-range B-52 strategic bombers on alert for Strategic Air Command.

KAZEL: *In the 1990s, you often said nuclear abolition was a good long-term goal, but we could greatly reduce risks and maybe avert catastrophe by de-alerting our land-based weapons – our ICBMs [Intercontinental Ballistic Missiles]. What do you think about 450 of our Minuteman missiles being on high alert nearly 20 years later?*

BUTLER: I just scratch my head in astonishment, quite frankly. That is such a feature of the Cold War era, when we were vulnerable to attack at any moment, so we had to be prepared to launch our forces on a moment's notice. Somehow, we've managed to persuade ourselves that we need to continue in that posture. I find that unseemly.

I don't know how many hundreds of billions of dollars of investments we have in Russia, and how many hundreds of thousands of our citizens are there at a given time. But somehow, just the thought of having an instantaneous circumstance in which we would even consider annihilating our own people, not to mention their assets, in Russia, boggles the mind.

From the standpoint of what I consider rationality, it doesn't pass muster with me. Even when you have someone who can be irresponsible in his rhetoric, like a [Russian President Vladimir] Putin, I find it extremely difficult to imagine a scenario, in this day and age, when you'd have a nuclear weapons component [to a US-Russian conflict]. A posture of immediate alert, I just think, is a complete anachronism at this point.

KAZEL: *As commander of US nuclear forces, you helped persuade the George H. W. Bush Administration to take our nuclear bombers off high alert. Did you or other military leaders also consider taking ICBMs off alert?*

BUTLER: The farthest we proceeded was to retire the Minuteman IIs early, and then move in the direction of single-warhead Minuteman IIIs. The community was just not ready for the prospect of taking two of the vaunted Triad weapon systems off alert at the same time. It was a pretty big bite to swallow just to get those bombers off alert. Part of the price of that was everything else was considered sacrosanct.

KAZEL: *Is there even a role for land-based missiles now? It seems they just increase danger by turning large portions of our territory into enemy targets.*

BUTLER: I would have removed land-based missiles from our arsenal a long time ago. I'd be happy to put that mission on the submarines. I came to develop an extremely high regard for submarines – their flexibility, their invulnerability, etc. And contrary to myth, we can communicate with them very quickly. So, with a significant fraction of bombers having a nuclear weapons capability that can be restored to alert very quickly, and with even a small complement of Trident submarines – with all those missiles and all those warheads on patrol – it's hard to imagine we couldn't get by. Now, the Air Force would take exception to that.

KAZEL: *How do present alert levels add to the pressure to respond quickly? What did those simulations tell you?*

BUTLER: The biggest constraint is how tight the timelines are: about 10 minutes for the conversation between [the commander of nuclear forces] and the President to characterise the attack and then get a decision. It's a function of physics. If you take a nominal ICBM launch from the former Soviet Union to a nominal target, say, in the centre of the United States, you're talking about 30 minutes. But that missile launch has to be detected and then its trajectory has to be assessed through two independent warning means, satellite and ground radar. Well, that takes time. It takes about 15 minutes or so, and then the commander of NORAD has to be brought on line. All the while, more and more missiles are being launched and are coming onto the radar screen. At some point, the commander of NORAD has to say, 'OK, I understand now what's happening, we're under nuclear attack'.

I would call the President. I had to lay out his options. The Nuclear War Plan has four major attack options with increasing numbers of weapons. Essentially all of our [recommendations were] MAO-4, Major Attack Option 4, which means we send everything we've got and it would cover all 10,000 weapons [in the U.S. arsenal at that time].

The war plan was my responsibility and had about 12,000 targets. Some weapons could destroy more than one target at the same time. That war plan was premised on all those weapons being available. If we allowed the Russian attack to take out any number of them, then the war plan may not have been viable. So, there reached a point where I would say, 'Mr. President, I need your decision now'. It became unequivocally obvious to me the immense responsibility of my role. The [person playing the role of the] President said at that point, 'General Butler, what is your recommendation?' And effectively, on my answer – which he would always endorse – hung the fate of civilization. If I said MAO-4 [full-scale counterattack], he would always agree. If it's you sitting in that chair, knowing that once you said the fatal words, 'I agree with your recommendation: MAO-4,' you've sealed the fate of hundreds of millions of people outright. Thanks to what science has now taught us, as Carl Sagan said many years ago (see *Spokesman* 46, 57), the combined effects of a wholesale attack involving 20,000 [American and Russian] thermonuclear warheads essentially would have been the end of civilization as we know it.

KAZEL: *When you commanded Strategic Air Command, you ordered a major revision of the list of nuclear targets. You saw that US missiles would hit political or military targets, but they'd be so numerous and so powerful that countless civilians in nearby cities and industrial areas would also be wiped out.*

Butler: The notion of the '1, 2, 3, 4' major attack options is that we would begin with leadership targets, MAO-1, then military targets, then military-industrial targets, and then, finally, urban areas. That was always parsing matters pretty finely, because if you think about where the leadership targets were, they were in Moscow, for the most part. We had 400 warheads targeted on Moscow, in my day. Think about that: four hundred thermonuclear weapons, each on the order of 350 kilotons? The urban population certainly was not going to be spared simply because it was an attack on military leadership.

Kazel: *Your book looks critically at US military leaders who kept using deterrence theory to justify large nuclear arsenals long after the Cold War ended. Do you think the new generation of leaders really still has faith in deterrence?*

Butler: In the latest issue of the Air Force's *Air and Space Power Journal,* there was an article on the five 'myths' of arguments against deterrence, one of which was that submarines can do the job. I don't want to use the word 'desperate', but I think it is a very concerted effort on the part of the Air Force to shore up deterrence. It all rings so hollow. It's like the priest in the temple, and his voice is echoing amongst the columns. But you peel it all back and say, 'Give me a situation when you would actually use one'.

Kazel: *Why isn't deterrence reliable?*

Butler: The first victim in a nuclear crisis is deterrence, because at that point all bets are off. One of the fatal flaws of deterrence, then and now, is deterrence applies in a circumstance in which you have a mortal enemy. If they're a mortal enemy, you probably don't have a very close relationship with them, which means you probably don't understand how they think to a very high degree of precision – how they will respond under very threatening circumstances. That doesn't sit very well with me. Even people who live with each other every day are sometimes hugely surprised.

I can't think of a situation where you're more likely to be wrong. And in the case of the Soviet Union, we were wrong. They didn't buy into 'flexible' response. When I was a young, so-called 'action officer' in the Pentagon and Air Force Headquarters, an Air Force brigadier [general] named Jasper Welch – who would shortly become my boss – had spent a very long period of his life developing a whole notion of 'flexible' options to be incorporated into the Nuclear War Plan so that the President wasn't left with an all-or-nothing [decision]. Now on the face of it, that seems like a very reasonable idea, instead of the President having to say, 'That's all I can do, just launch everything?' What if it's not an all-out attack [on the US]? That was perfectly rational and legitimate: limited and selected attack options.

Kazel: *As I understand this, our government started to think it might selectively strike Soviet political and military targets in a war, but that launching a massive attack at the beginning might be avoided?*

Butler: This is one of the frailties of deterrence: the Soviets never bought into that sort of philosophy. For them, one warhead on the motherland, and that was it. If they should lose their leadership in an initial attack, they made provisions for the forces to be launched [by various lesser officials or military officers]. So, that gave the lie to the whole notion of flexible options.

If I could strike one word from the lexicon of the nuclear weapons enterprise, it would be 'deterrence'. Because it's easy. It's lazy. It's using rhetoric for a replacement of really rigorous thinking about what is exactly implied by your actions. I would force people to actually describe what it is they think they're doing [by holding onto nuclear weapons] in very detailed terms, and then defend it on that basis.

Kazel: *You've also said that deterrence fails in practice because it ignites arms races – spiraling costs, new and revamped technologies, and instability as various nations seek not just security but superiority. Can deterrence theory give birth to conditions that are likelier to lead to war?*

Butler: We had 40 years of history of that. Of the several things that deterrence did not do, it did not serve as any sort of guide for force levels. To the contrary, in service to deterrence, force levels constantly increased almost exponentially. At its height we had 36,000 weapons in our active inventory. Imagine that. Thirty-six thousand. We had warheads on artillery shells that could be launched from jeeps. We had dozens of warhead types and delivery systems. We had landmines and sea mines. We deluded ourselves with regard to our capacity to manage crises. Deterrence was an open-ended invitation to just build more, and more complicated, systems. We always found a reason to go to the next level of arms, or the next hot technological biscuit to come popping out of the oven. That just scared the bejesus out of the Soviets. They weren't equipped to stay in that race, and ultimately Ronald Reagan saw that and he spent them into oblivion. The weapons became ever more numerous and more destructive. There's no end to it, and the rationale was always that deterrence required it.

That we would pursue these weapons with such unfettered enthusiasm – competing amongst the [military] services for resources, going to roll out these shiny new things, cutting ribbons – spoke to me a great deal about the human condition. For some people, technology has absolutely mesmerising qualities. If we can do it, we must. That's the kind of thinking that got us where we are today. To add to that, the military-industrial complex was being fed a virtually endless trough of money. There is no end to the number of people who will find any way to justify building something new, brighter and better. I've seen that happen time and time and time again …

Fifteen hundred nuclear warheads [deployed by both the US and Russia today] is still a mind-boggling amount of destructive potential. Mind-boggling. I can't think of anything that underscores that better than how concerned we are about one falling into the wrong hands. We still readily accept 1,500 as a reasonable number. That's the kind of 'logic' that we get locked into, in the nuclear era.

Kazel: *Aside from questions of strategy, you argue in your memoirs the same position you pounded home 20 years ago: that nuclear weapons violate basic ethics and are an affront to sacred human values. How did you start to believe that?*

Butler: I guess I'm what some would call a spiritual person, from my study of the origin and evolution of the universe and life as we know it. I find it so awe-inspiring. That's where I draw my sense of morality and humanity, and the overwhelming importance of sustaining the privilege of life. I think our no. 1 responsibility as human beings is to continue to elevate the bar of civilised behaviour, to make conditions [hospitable to] life, liberty and the pursuit of happiness. That's what captures my spiritual fancy these days.

Rather than being concerned about the moral implications of these devices, we continue to pursue them as if they were our salvation – as opposed to the prospective engine of our utter destruction. Human beings are by far the most destructive species on the planet; that the planet has ever seen. We kill each other for a variety of reasons, ranging from pleasure to vengefulness to fear for survival. As long as these weapons exist, and people hold them in such high regard for reasons of national esteem, they act as a brake on our capacity for advancing our humanity.

Kazel: *Were you already thinking that when you were commander of Strategic Air Command?*

Butler: As they say, the prospect of hanging wonderfully concentrates the mind. When I finally began to understand the full import of the rather terse phrase 'weapons of mass destruction' – their acquisition, their operation, their targeting, the execution of that war plan – I began to think and reflect more deeply on the question, how did we ever get ourselves into this circumstance? And, by extension, how are we ever going to get ourselves out?

We had become so enamoured of nuclear weapons. Somehow, they had become an end in themselves; sort of a totem of power.

People embark on these sorts of messianic quests. Some people climb Mount Everest, some want to swim the Amazon. But for me, it was more about telling the story of how we escaped the Cold War without a nuclear holocaust by some combination of skill, luck and divine intervention – probably the latter in greatest proportion, for those who do believe in religious intervention. Because skill and luck certainly don't account for it.

Kazel: *Do you believe it's ever ethical to bomb civilians in wartime?*

Butler: (Pauses.) One of the very first victims of all-out war is ethical considerations. In World War Two, as an example, we burned 200 of the largest Japanese cities to the ground. In the history of warfare, circumstances have produced a sort of scorched-earth policy and ethical considerations are simply abandoned. That was certainly the case in Hiroshima and Nagasaki. There were ostensibly military installations nearby, which were used to rationalise the bombings, but there were other targets that we could have struck that would have put far fewer civilians at risk.

The cold, hard fact of the matter is that a nuclear weapon is, at its very core, anti-ethical. It is simply a device for causing wholesale destruction. Nuclear conflict is essentially an irrational activity, because essentially what you're doing is signing your own death notice.

Robert Kazel is a Chicago-based writer and was a participant in the 2012 NAPF Peace Leadership Workshop. This entry was posted in Events, Nuclear Abolition and tagged Lee Butler on May 27, 2015 by Robert Kazel.

www.wagingpeace.org

More than 300 people attended the inaugural Ken Coates Memorial Lecture at Nottingham University on 3 June 2015.

We publish Frances O'Grady's keynote address, together with two introductions. Andreas Bieler, Professor of Political Economy at Nottingham University, chaired the meeting, which was, by general acknowledgement, a memorable and meaningful occasion in the tradition of labour movement activism which Ken Coates encouraged.

On 16 September 2015, a follow-up seminar, led by Dexter Whitfield, will consider 'Alternatives to privatising public services: Capitalist dynamics reconfiguring the state'.

The Ken Coates Memorial Lecture is sponsored by the Bertrand Russell Peace Foundation and hosted by the Centre for the Study of Social and Global Justice, (University of Nottingham) with support of the local University and College Union association, Unite the Union East Midlands Region, other trade unions, and Five Leaves Bookshop.

Educate, Agitate, Organise!

Frances O'Grady

Frances O'Grady, General Secretary of the Trade Union Congress, gave the inaugural Ken Coates Memorial Lecture at Nottingham University on 3 June 2015.

I've been invited to talk about the future for the Left and the Labour Movement. I have to admit that when I accepted the invitation I'd hoped things would be looking a little more optimistic in Britain. Then on Election night at 10pm I saw the exit polls. All the opinion polls had pointed to a Labour-led coalition. Paddy Ashdown famously promised to eat his hat. And even the institution I'd come to trust most, the other Paddy, Paddy Power, miscalculated the odds. So there were plenty of people left scratching their heads about what went wrong.

Luckily, before the week was over, Tony Blair popped up on our TV screens to give us the answer. Apparently, Labour lost because it was too left wing. In my view, he was wrong: any Labour manifesto worthy of the name should be about building homes, raising wages, and protecting the NHS. That's what I'd call real aspiration.

In my view, if anything, Labour's problem was its failure to counter the Tory mantra that the deficit was caused by too much spending on public services early enough. And the admission that the previous Labour government had been too soft on banking regulation came far too late. That view is supported by research commissioned by the TUC and carried out straight after the election, which was published two weeks ago.

Far from thinking Labour was business bashing, voters believed that controls needed to be tougher. Labour was seen as being on the side of ordinary people and was most trusted with the NHS. But views on Labour's economic competence, and on welfare and immigration, were all unfavourable – sometimes brutally so.

Another key factor in the last weeks of the campaign was the Conservative Party's shameless playing of the nationalism card, whipping up fear of a minority Labour government dependent on SNP MPs. Our evidence shows that it worked. If the SNP slogan was 'put Scotland first', then it followed that people in England would worry that they would be put second. It is one reason why I believe that, while the level of participation and debate in Scotland may be exciting, the politics of place is ultimately an inadequate response to neoliberalism. Fragmentation on the left always favours the right. And I've long believed that working people will always have more in common with each other than we will ever have with a London stockbroker or, for that matter, an Edinburgh banker.

But I want to say a few words about Ken Coates and his legacy. Our paths briefly crossed when Regan Scott commissioned Ken and Tony Topham to write the history of the Transport & General Workers' Union. I'd started work at the T&G at the old head office in Smith Square when Ron Todd was general secretary. I was in awe of the union's rich tradition of working class intellectualism that combined a commitment to socialism with industrial militancy.

This wasn't a purely theoretical exercise. It was borne of real experience. For me, Ken was the embodiment of that tradition of intelligent trade unionism and practical socialism. It's no exaggeration to say that Ken was one of the most perceptive and eloquent figures on the Left in post-war Britain. Other names may have hogged more headlines, but few could match his immense body of work.

Ken's political career began at the age of 15, when he won a mock election at Worthing Grammar School. Perhaps the first and only time a Communist has topped a ballot in West Sussex. From that moment on, Ken dedicated his life to the cause of organised labour. As a miner he understood what hard, dangerous, manual work was like; but also the spirit of solidarity that can flow from it. As a teacher and academic, he was deeply committed to the cause of adult and trade union education. As a writer, he famously charted how poverty corroded working class lives here in Nottingham. And as a socialist he firmly believed in public ownership – fighting tooth and nail New Labour's decision to ditch Clause 4.

Just as fundamentally – as befits someone deeply involved in Bertrand Russell's Peace Foundation – Ken was a committed internationalist. He became a miner because he refused to fight the communist uprising in Malaya after the war. He resisted American militarism in Vietnam. He opposed Soviet intervention in countries like Hungary and the former

Yugoslavia. And he was at the heart of the European movement for nuclear disarmament.

In many ways, Ken's life reflected the ebbs and flows of the Left, taking in as it did the fortunes of communism, old and new Labour, social and neoliberal Europe. I want to draw a thread from Ken's life and work, to the future. However, while I'm a great believer in learning from labour history, equally, I refuse to be a prisoner of it. Perhaps the Left needs a little less nostalgia for a supposed golden bygone age and a little more hardheaded analysis to help us shape our future. In particular, we need to address how capitalism has changed and how that means labour must change too.

We are all familiar with the trends: growing inequality with top bosses' pay rising by 26% since the crash while everyone else's wages have tanked; the role of new technology – not just in the workplace but in our home lives too, with workers increasingly expected to work for free, from administering our own bank accounts to the nightmare of assembling flat pack furniture; globalisation and how that feeds nationalism and political disaffection; and the fact that just a handful of multinational corporations – many of them investment banks and private equity groups – now dominate corporate wealth.

But we are less practised at working out what all this means for the battle of ideas, for social change and for industrial power. And, more to the point, what we need to do about it. Will the next wave of unionisation come from the new global middle class? Or from the great masses of the precarious poor? Where will the new front lines of that battle be drawn? And what do new business models mean for collective bargaining, still one of the most effective ways of redistributing, not only wealth but also power. As an experienced union organiser once told me, there is no such thing as an unorganised workplace. It's just the employer got there first. In other words, our job is to match and then surpass them.

I want to highlight three of Ken's great interests and raise questions about what this means for the wider labour movement over the next period ahead. The first is Ken's pioneering work on workers' voice, through the launch in the 1960s of the Institute for Workers' Control; an influential group that captured the support of large numbers of shop stewards, as well as trade union leaders such as Jack Jones and Hugh Scanlon. Ken's focus on achieving reforms to give workers a role in running their enterprises was nothing if not forward looking.

It paved the way for the Bullock Report on industrial democracy a decade later, to my mind, despite Jack Jones' best efforts, one of the great missed opportunities for the British Left and the then Labour government.

But perhaps more fundamentally, it inspired working people to take charge of their working lives, generating a multitude of models; workers' co-operatives like those at Meriden, Kirkby and the Scottish Daily News; workers taking control of their plants, such as the Upper Clyde Shipbuilders work-in; and so-called 'workers' plans' for alternative production to meet social needs.

The best known of these plans was the Lucas Aerospace shop stewards' proposal in the face of plant closures. They worked up detailed plans for switching from arms production to a range of 'socially useful products', from portable kidney machines to hybrid road/rail buses. The industrial history of the 1970s is too often reduced to lazy clichés about mindless militancy and union dinosaurs. But the movement for workers' control, of which Ken was such an integral part, presents a very different story.

I still have a copy of the workers' plan for the motor industry, produced by shop stewards and sympathetic academics in the 1970s. My dad was a steward when he worked on the line at Cowley for British Leyland – a company whose industrial relations were characterised as the ultimate basket case. Yet on page 87 of that workers' report you will find a cogent environmental and industrial case for investment in the development of electric cars. The stewards were all too aware of the challenge of climate change, that the British car industry was in trouble, and that prototypes were already being developed in Japan. Sadly, management and the government rejected the plan. It wasn't until thirty years later that a BMW electric mini finally rolled off the line at Cowley.

Imagine how different things would have been if management had listened to workers and their unions all that time ago. This is one reason why – and I recognise this is a long-term project – the TUC has been campaigning for representation of workers on top pay committees and, indeed, on boards. Our boardrooms are badly in need of the ingenuity, honesty and plain common sense union reps can bring. Not as an end in itself but as the beginning of a journey of economic democratisation.

The propaganda that entrepreneurs are the sole wealth creators has been peddled for so long that even some of those politicians who should know better seem to swallow it. But it's workers who create wealth and they should be entitled to a voice in the fair distribution of it. Similarly, we are saddled with a corporate governance system premised on the notion that shareholders – who hold shares, on average for just a matter of months – are the best stewards of the interests of a company. Yet nobody has a stronger interest in the long-term success of a company than those whose livelihoods depend on it.

Now, in the shadow of the election result, I realise that, to some, the TUC's campaign for workers' voice may sound fanciful. But surely no more so than how it must have felt for those who began the long campaign for political suffrage for working people centuries ago?

And, of course, today the majority of European Union states already guarantee worker representation in some form, up to and including board level.

Which brings me to another of Ken's preoccupations that still has relevance today – a deep commitment to social Europe. During his ten years as a Member of the European Parliament, Ken earned genuine respect for speaking up for the interests of ordinary workers, pensioners, the disabled, and the unemployed – if not always from the party he represented. He was chair of the European Parliament's human rights sub-committee at a time when the continent was faced with the terrible conflicts in the Balkans. He worked with Jacques Delors and Stuart Holland to promote not just full, but decent, employment; an ideal the current generation of EU leaders would do well to remember. And he argued, convincingly, that without a strong social dimension, that bargain between capital and labour, popular support for the European project would wither on the vine.

Since before the crash, the European Commission, along with the International Monetary Fund and European Central Bank, has pursued an agenda of austerity, liberalisation and deregulation – including taking a sledge hammer to collective bargaining in the programme countries. So the political class should not be surprised if workers around Europe start to wonder - what's in it for them?

Of course, David Cameron's gamble of an EU referendum is premised on the belief that workers will buy curbs on migrants' rights to benefits as the answer to their problems. Never ones to miss an opportunity, some business leaders are lobbying to get workers' rights, particularly those for agency workers and on working time, together with a moratorium on any new rights, thrown into the 'reform' package. We are told Europe must become a jobs factory. But without workers' rights it would become a sweatshop.

The government is playing with fire. Far right parties take many different forms across Europe, but they have one thing in common: they all thrive on scapegoating migrants. And making exploitation even easier, for example through employing migrant workers on inferior agency contracts, which undercut local workers and union agreements, will only add fuel to the fire.

With the EU referendum fast approaching, last week I issued a warning through the pages of the *Financial Times*. Our message was simple: you will not win blue collar votes to stay in the EU by worsening workers' rights. As Ken himself would have been the first to recognise, the challenge for us on the Left now is to restate the overwhelming case for a Europe for people not just corporations – and the progress of that trade agreement, TTIP (Transatlantic Trade and Investment Partnership), will be one key test of that. It means taking the fight to those who want to get rid of all those rights – to paid holidays, to consultation at work, to health and safety – that emanate from Europe. And it means pushing back and framing the referendum debate along positive, progressive lines.

Of course, we in the trade union movement will be having our own debate about the referendum at our Congress. I hope it will set out our alternative vision for the European Union; one that is fit for the twenty-first century, and sets new standards across the EU, including for collective bargaining and economic citizenship; a Europe of investment for good greener jobs and decent wages; strong rights, and corporate governance reform for a voice at work; tough regulation of excessive corporate and financial power; and good public services fit to meet modern needs, such as childcare and care for the elderly. And, last but not least, a transactions tax (the 'Robin Hood' tax) on the finance sector, to help pay for them.

The third area to consider is union organisation. Ken and Tony Topham's book on workers' control also included a brief history of the rise of mass trade unionism. The book was titled *The New Unionism*, and I bought a copy in 1974. This work had a big and lasting impact on me. So much so that I called the project that launched the TUC's Organising Academy, 'New Unionism'. I admit this was in part designed to mirror – some might say gently mock – New Labour. But I also wanted our young trainee organisers to learn about, and feel inspired by, that first wave of mass trade unionism, including it's small band of paid union officers who combined the roles of organiser, agitator and educator.

It is no accident, all the challenges our country faces: low productivity borne of low investment and low wages; lousy jobs that frustrate the talent of our young people; and growing inequality whereby the five richest families in the UK have more wealth between them than the poorest 12 million …

It's no accident that the new Tory government's top priority is to impose new thresholds on union strike ballots. Not only to impose thresholds but also to give employers new powers to break strikes with agency workers, and potentially criminalise pickets and make them the subject of new state surveillance. I often explain this attack by highlighting the fact that unions are the last line of defence against the colossal cuts in the pipeline. But I think it goes even deeper than that. Neoliberalism and extremely unequal societies necessitate more authoritarian models of democracy. Trade union values and organisation – still six million strong – act as a living, breathing and sometimes rebellious rebuke to all that.

But, while this government may be ruthless, they know that they are not invincible. It is, after all, a weak administration with a much smaller majority than the previous coalition government. If UKIP can mount a rebellion against its leader with just one MP, think what those Eurosceptic Tories can do. And David Cameron has already said that, in the interests of his country – or was it his career? – he will not stand for another term. That means that the contest for the next leader of the Conservative Party has already started and it will last a full five years. In fact, the only face on election night that looked more disappointed than Nigel Farage was that of Boris Johnson.

I want to make one thing clear: if the Conservatives think we're going to give up our freedom of speech; give up and go quietly; surrender the human right to protest against workplace injustice – then they better think again. But, without doubt, we have our work cut out to defend and rebuild the movement. We will have to box clever and re-learn the art of broad alliance building; something that came as second nature to Ken Coates.

There are lessons to learn from around the world; the mobilisation of young people into a mass movement for change in Spain; the fast food workers' campaign for a living wage in the United States; the bus drivers' action in Brazil.

Closer to home, we need to champion communities that have long felt abandoned. A new activism is emerging on the grassroots Left, from anti-austerity organisations such as 38 Degrees to green campaigns such as Reclaim the Power to debating groups such as the Post-Crash Economics Society. But we need to live our values in real communities, not just virtual ones; take Focus E15, a housing protest group. It began when housing benefit cuts led a London council to try and move single mothers out of a hostel to cheaper private rented accommodation, 100 miles away. Like many community campaigns, it's driven by young people – including young working-class women. Those young housing campaigners do not necessarily see traditional party politics as the answer, nor what the former TUC leader, Walter Citrine, once called 'the traditional walls of trade union policy'. But they are often highly skilled networkers and campaigners.

What if those skills were put to work in union organising? Fewer than one in five workers in the private sector now hold a union card. For young people the figure is even lower. Yet young people are on the front line of flexibility, which means zero hours, low pay and no prospects. Organising isn't easy. Where once life seemed simple – Labour against the Tories, workers against the bosses, underpinned by strong class identity – building the movement today is more complicated. But for all we are living through a state of political flux, remember our movement has shown itself to be resilient throughout our history.

Together we must show we've got intelligent answers to the central challenges of our time. How we build a greener, fairer economy with decent jobs, wages and homes at its heart. How we safeguard the public services and social security that are the hallmarks of any civilised society. And how we address those profound inequalities of wealth, income and opportunity in modern Britain.

This parliament will see the TUC launch a major new initiative to reach young workers – the baristas, the shop workers, the cinema workers. We aim to help them self-organise in their workplaces, with the backing of families, communities and the general public. As the experience of last year's dispute at the Ritzy Cinema in London showed, this is a generation with a lot of courage and little to lose. Those workers led 13 strikes and won a 12 per cent pay rise in their campaign for a living wage. A great achievement. And one that confounds the stereotype of young people as apolitical and apathetic. There's plenty we can all learn from how they did it: from musical picket lines to Facebook campaigns to winning public support and shaming the company.

But if we are to succeed, let's not pretend that trade union structures and organisations can stay the same. Capital has changed and so must we. At a time when trade union resources are under pressure, we have to think hard about how we empower workers to represent themselves. In the long run, I've always believed that our destiny, as workers, is in our own hands. From Mary Macarthur to Bob Crow, from Jim Larkin to Jack Jones, the trade union movement has always been at its best when we have educated, agitated and organised. Together we must address the growing disconnect many ordinary working people feel with politics, work and life in general. And if we do not, then we can be sure about one thing: others will.

We've already had a glimpse of what can happen when nationalists offer people easy solutions, appealing to romantic but ultimately false new dawns. And we know how in an age of anxiety, when people feel scared, insecure and powerless, they can turn on the vulnerable and disadvantaged. That's what's at stake when we talk about the future of our Labour Movement.

With the Tories back in power, these are undeniably tough times for us. But while this is a moment for hard-headed analysis, there are grounds for hope, too. If we remain true to our values of equality and solidarity; if we find new ways to reach out to new groups of workers, supporters and campaigners; if we resist the temptations of factionalism and stick together in unity; the Left, the Labour Movement, working people – our fortunes rise and fall as one.

Because the struggle for equality, fairness and justice to which Ken Coates devoted his life remains as urgent and as vivid as ever. Labour can't do it on its own. Trade unions can't do it on their own. Social movement campaigners can't do it on their own. But here's the point.

We can do it together. And if we join together and organise together, then together we will win.

For audio and video of the lecture:
http://www.nottingham.ac.uk/cssgj/centre-activities/previous-events/ken-coates-memorial-lecture.aspx

Frances O'Grady at Nottingham University

Better Idea

Tony Simpson

The author worked with Ken Coates from 1980 until 2010. He focuses on industrial democracy in his brief introduction to the Ken Coates Memorial Lecture.

Picture the scene. A windowless room. Earphones on. Suddenly, Jeremy Paxman crackles across the airwaves:

> 'What's so precious about your conscience, Ken Coates?'
>
> 'It's as precious as yours, Jeremy!'

comes the swift reply.

It was 1994, and Tony Blair wanted to ditch the Labour Party's constitutional commitment to common ownership, stated in Clause 4 of the Party's constitution. Ken with some other Labour Members of the European Parliament opposed Blair's changes. Although he had misgivings about the existing clause, Ken had much deeper concerns about the direction in which New Labour was going.

So it was, some 16 years later, in May 2010, that Ken wrote an encouraging letter to Ed Miliband, who had just declared that he was standing for the leadership of the Labour Party. Ken wished Ed well, adding that he was no longer a member of the Labour Party himself, having been expelled by New Labour, and continued:

> 'Clause IV was a highly imperfect document, much aggravated by the false identification of common ownership with Morrisonian forms of nationalisation … That part of the Clause IV commitment that was vital was to the best obtainable form of democratic administration.'

Extending democracy, particularly at work, was the main motivating idea of the Institute for Workers' Control, whose banner was raised aloft by Ken and other comrades (see *Spokesman 128* cover). Tamara, Ken's wife, remembers making the banner for Levellers Day at Burford in 1976. Incidentally, the IWC held a well-attended weekend conference in this very building at

Nottingham University, some months after Jim Callaghan lost the 1979 election. Tony Benn, Ron Todd and Eric Heffer all spoke in the room next door. One of the main themes, in addition to the early signs of Mrs Thatcher 'de-industrialising' the economy, was democratisation of the Labour Party. Those constitutional changes, initiated in part by Rushcliffe Constituency Labour Party, of which Ken was secretary, widened the franchise for electing the leader of the Labour Party to include trade unions and constituency labour parties. Previously, the electorate was confined to a few hundred Labour MPs.

A principal reason in my mind for inviting Frances O'Grady to give the inaugural Ken Coates Memorial Lecture is her consistent advocacy of 'workplace democracy' and the need for workers' voices to be heard in corporate governance. We are very grateful to her for making the time to come and talk to us in the early days of a new government, which has hastened to announce changes that will make it more difficult for trade unions to fulfil their democratic role and basic purpose of protecting and advancing rights, wages and conditions at work.

Jack Jones was held in high regard by Ken Coates, and Frances has put on record her own appreciation of Jack's encouragement and his 'intelligent trade unionism'. Jack Jones was elected General Secretary of the Transport and General Workers' Union in 1968 and continued in post for an eventful decade. He 'sustained a programme of demands for industrial democracy which culminated in the Bullock Report' (of 1977),' was the way Ken Coates described events to Ed Miliband in his letter. Ken was concerned about slavery at the workplace, which

> 'has created a great mass of people whose status accustoms them, while at work, to taking orders, rather than shaping, or helping to shape decisions … to reopen the argument about industrial democracy, as widely as possible, is a crucial choice, which has become more relevant with the growth of new technologies, and with the extensive polarisation which has given birth to our underclass. The servile industrial state generates a dire state of mind … it will surely be impossible to create a self-confident Labour Movement which remains in thrall to this mindset. You need to rehabilitate the goal of freedom in order to revive any tolerable aspirations to democratic socialism.'

Democracy and freedom were key elements of Ken's 'better idea', as he called it. Otherwise, he said, 'the helots will continue to vote for helotry,' which is one way, perhaps, of looking at recent events.

Ken Fleet and Regan Scott are with us tonight. Over decades, they worked with Ken on campaigns advanced by the Institute for Workers' Control, those organised by the Bertrand Russell Peace Foundation, such as European Nuclear Disarmament, and on sustained work for full employment in Europe, whilst Ken was an elected member of the European Parliament from 1989 to 1999. Stuart Holland continues that work with his sustained advocacy of a New Deal for Europe, persuasively argued in his recent book, *Europe in Question – and what to do about it*.

Sadly, Michael Barratt Brown, the founder of Northern College for adult students, the 'Ruskin of the North, is not with us. Michael died in May, in his 98th year. You will find some of the books he co-authored with Ken in the list Abi Rhodes has prepared, together with those written in collaboration with Tony Topham, Ken's long-term comrade in the IWC and other campaigns, who died in 2004.

We in the Bertrand Russell Peace Foundation, and Tamara, as well, are grateful to Tom Unterrainer, activist and teacher, and Ross Bradshaw, writer, publisher and bookseller, founder of Five Leaves, Nottingham's thriving radical bookshop, for seeding the idea of the Ken Coates Memorial Lecture and helping to make it happen. We also acknowledge with gratitude the generous help and support of Professor Andreas Bieler of Nottingham University's Centre for the Study of Social and Global Justice for hosting our meeting here, and that of the local University and College Union association and UNITE the Union in the East Midlands.

Tom is preparing an annotated bibliography of Ken Coates' extensive writings, which will supplement the archive of his mainly Parliamentary papers, already held in Nottingham University's Manuscripts and Special Collections. As well as graduating from this University, as he describes in *Spokesman 116*, entitled *Resist Much, Obey Little*, Ken Coates also worked in its Adult Education Department for some 30 years, where he taught day release classes for trade unionists, including many miners, workers from Rolls Royce in Derby, and local government workers. In addition, he taught evening classes at the Workers' Educational Association in Shakespeare Street, which is where some of us first met him. In collaboration with Richard Silburn, who is here, that gave rise to the St. Ann's survey and, eventually, to the Penguin Classic, *Poverty: The Forgotten Englishmen.*

Tom Unterrainer's work in progress will form part of the continuing programme to make full use of the rich legacy of Ken Coates' life and work for democracy, freedom and peace.

'to secure for the workers by hand or by brain the full fruits of their industry and the most equitable distribution thereof that may be possible upon the basis of the common ownership of the means of production, distribution and exchange, and the best obtainable system of popular administration and control of each industry or service.' (Clause 4).

Discovery

Tom Unterrainer

The author is compiling a bibliography of Ken Coates' many writings. This brief introduction to the Memorial Lecture is based on his article 'Ken Coates and socialist renewal', which appeared in Spokesman 128.

In October 2012, I 'discovered' Ken Coates. I say 'discovered', but in fact I'd heard of Ken more than ten years before, in connection with anti-war campaigning. What I really mean is that I started to read and think about Ken's very many books, articles and his political activities.

The circumstances of this discovery are worth re-telling, if only because they demonstrate the reach and enduring influence of Ken's ideas. Leo Panitch, the Canadian political economist, visited Nottingham to deliver a lecture on his book *The Making of Global Capitalism*. At the opening of the lecture, he asked: 'who in the audience has heard of Ken Coates?' Despite knowing the name, I didn't raise my hand. Neither did anyone else in the room.

Panitch went on to explain that the last time he visited Nottingham was in the 1970s whilst completing his PhD and the sole purpose of the trip was to meet Ken. On this recommendation, I ordered a copy of Ken's collection of essays, *Beyond Wage Slavery,* and was captivated. Two things impressed me about the essays. Firstly, the constant challenges to the reader: 'what next?', 'if we can't do this what can we do?', 'why not?' These essays are no meander around some aspect of history or theory; they are sharply written expositions with a practical purpose in mind. Secondly, the range of ideas covered in *Beyond Wage Slavery* spoke of a socialist actively engaged in a wide range of issues.

First and foremost, through these essays Ken is addressing himself to the Labour Movement. The purpose of his writing is not to mobilise some lacuna of collective thought and compliance but to challenge and organise activity in the movement as a whole. As I was to discover, from the early 1960s onwards Ken was involved in efforts to mobilise, organise and educate the Labour Movement on a socialist basis.

Throughout the 1960s, Ken and a growing band of comrades worked to make socialist ideas relevant to the realities of the movement in Britain, launching sharp interventions into the living and breathing organisations of this movement. Ken's considered attack on the Wilson Government's supplicant relationship with the United States over its war on Vietnam was not simply the outraged response of a morally offended individual, but the expression of a growing movement of discontent. The Institute for Workers' Control was not a political front marshalling imaginary battalions, but an organic expression of a substantial movement within the trade unions and Labour Movement more widely.

Ken's diverse activities were guided by a perspective that was rooted in the working class movement, informed by international events and concerns, inspired by contemporary intellectual developments, but driven by the imperative to make ideas a reality. He does not provide all of the answers to all of the questions we now face, but Ken's work – his voluminous writings and political initiatives – is a vital reference point for those of us determined to build and renew the socialist movement in Britain. It is a vital reference point for those of us who see the connections between the continuing drive to war, environmental degradation, economic crisis, social cruelty, and political dishonesty.

Ken Coates' work is important because it demonstrates the historical existence of 'another socialism': a socialism rooted in practical reality, informed by international concerns and a sense of moral urgency. The work of renewing socialism can find no more powerful reference point than this.

* * *

Professor Andreas Bieler's account of the inaugural Ken Coates Memorial Lecture is available online, together with other useful links to his work on trade unions and global restructuring:
http://andreasbieler.blogspot.co.uk/2015/06/the-future-of-left-where-next-for.html

Michael Barratt Brown died on 7 May 2015, in his 98th year. He had a fall in early March, and never left hospital. Until then, he was in regular contact with us by e-mail, and had arranged to meet The Spokesman at his home in London. When we rang the bell, there was no reply.

Michael was our friend and comrade during many years, and was always ready to contribute to the work of the Russell Foundation. He was good company and great fun. We shall miss him very much. We publish tributes by Malcolm Ball, Victoria Bawtree and Stan Newens, three people who knew Michael well. But we begin with Michael's own words, as he explains how he met Ken Coates and sought to build 'solidarity and co-operative activity from below'.

'Yer not coming for me' Michael 2005 (Seekers)

Solidarity from below

Michael Barratt Brown

In his autobiography, Seekers, *Michael recounted how he met Ken Coates.*

… Side by side with my work in the extra-mural department [of Sheffield University], I became involved in founding and chairing a Society of Industrial Tutors. This was not a trade union but a think-tank, preparing materials and organising conferences of tutors from universities, trade unions, the Workers' Educational Association and technical colleges, all engaged in industrial studies courses. I had strong support from my deputy, Geoffrey Stuttard of London University. Geoffrey had, as I discovered rather surprisingly, been on the other side in Yugoslavia, supporting the Chetniks against Tito, but we worked well together. We even saw Michael Foot, the minister responsible, to persuade him to extend more widely shop stewards' day release for industrial studies. Michael had given a very generous review to a book which I had written called, rather hopefully, *From Labourism to Socialism*. The book was full of warnings about the concentration of capital in the hands of a few giant transnational companies, a fact which Michael recognised as of great importance. The response to this development by Labour at the top was not clear. We had to begin from below.

A further extension of my work with the industrial tutors was my involvement in discussions, led by Ken Coates of Nottingham University extra-mural department, on workers' control in industry. We had a first meeting in Nottingham with a number of trade union officials and several industrial tutors including Tony Topham of Hull University. Ken and Tony went on to write the major two-volume history of the Transport and General Workers' Union (TGWU). The Nottingham meeting initiated a long and close personal friendship between Ken and me, which ended only with Ken's death in 2010.

We formed an Institute for Workers' Control and organised a series of conferences with the support of leading trade unionists, in particular, Jack Jones of the Transport and General Workers' Union and Hugh Scanlon of the Amalgamated Engineering Union (AEU). Several thousand men and women attended our conferences in different cities, and we had the support of Tony Benn at the Department of Industry. He came to some of our conferences and took the concept of workers' control into actual practice in several industrial experiments. The involvement of Tony in our work led to the development of a very close relationship between him and Ken Coates, in which I had a very minor share. The Institute published a regular Bulletin, three Annual *Trade Union Registers*, and a good number of pamphlets and books on particular industries.

The climax came with the public ownership under workers' control of the Upper Clyde Shipyard. Three of my friends were involved in different ways, with me representing the Institute. Tony Benn gave industrial ministerial support. Robin Murray wrote up the story in a little book for the Institute, and Ken Alexander was appointed chairman of the company that took over the whole operation of the yard. One source of constant support was Michael Meacher MP, who became a minister for the environment in a later Labour Government.

Founding these organisations – industrial day release classes, an Industrial Tutors' Society, an Institute for Workers' Control – might just seem to be making space where one could talk and write together with agreeable comrades, and forget that the aim was to help them to lay the foundations for building a better society for working people. When in the next chapter we look at the founding of the Northern College and in the last chapter at the founding of a fair trade movement in the UK, the same question has to be asked about the long-term aim. It was never for me a top-downward exercise, but always an attempt to support the building of solidarity and co-operative activity from below.

THERNCOLLEGE

Northern College

Malcolm Ball

Michael Barratt Brown will, perhaps, be most readily remembered for his role in founding and establishing The Northern College. There is little doubt that this event has made a considerable impact on the lives of working people in the North of England and stands as a beacon both of what is necessary and what is possible. However, before his time as Principal of Northern College and for many years after he had left the post, Michael continued to influence the world around him.

Michael was born in Oxford on 15 March 1918. His education included the Dragon School in Oxford, Bootham School in York, and Corpus Christi College, Oxford. Following his formal education he served in the Friends Ambulance Unit stationed in London, 1940-41, and stationed in the Middle East, 1942-43. He moved to the Middle East Relief and Refugee Administration in 1944. From 1944 to 1947 he was with the United Nations Relief and Rehabilitation Administration (UNRRA). He served as Chief of Mission in Sarajevo and as Executive Assistant to Chief of Mission in Belgrade. It was during his time with UNRRA in Yugoslavia that he met his second wife and partner, Eleanor Singer.

Following the war, in 1947, Michael and Eleanor returned to England. They married in 1948 and settled in Colchester, following Michael's divorce from Frances, his first wife. Michael, although an economist, worked in the film industry as Assistant Director with DATA Films, 1948-1950, before moving into teaching as a part-time tutor with the Workers' Educational Association, 1951-1954. He took up a permanent full-time post, as lecturer, at

The author was a Derbyshire miner and day-release student and, later, a colleague of Michael Barratt Brown in founding Northern College, Barnsley.

◀ Michael and The Northern College, painted by Pat O'Neill

Cambridge University Extra-mural Department in 1955. It was also during this time that he met and befriended a young printer, William (Bill) Hampton, with whom I was to follow the route of adult student into adult teaching. Michael continued his academic career in adult education with a move to Sheffield University Extra-mural Department (EMD) in 1959 and remained there until 1976, when he left to take up the post as Principal of Northern College (See *The Northern College*, pp.36-69, 2004).

The period working at Sheffield Extra-mural Department was a pivotal period in Michael's career and in his political life. As a teacher of adults, he found himself working with the National Union of Mineworkers in Derbyshire and in Yorkshire, and with the steel industry trades unions. Active members of these unions were able to attend day release courses for up to thirty days each year for three years and the courses were taught by Michael and his colleagues in the Department. This was supplemented with a weekend residential course held at the Coal Industry Social Welfare Organisation (CISWO) holiday camp at Skegness each Easter. As a result of this adult education, large numbers of members moved into further education, including from the pit to university, via the then existing adult residential colleges in Harlech, London and Oxford. One important consequence has been that there are many local politicians and teachers and local state administrators who have passed through the hands of Sheffield Extra-mural Department. The relationship resulted in many friendships with former students and with the union leaderships. Herbert (Bert) Wynn, the General Secretary of the Derbyshire Area of the National Union of Mineworkers, co-wrote a campaigning booklet, *A Plan for the Miners*, in 1964. Michael assisted with the economics research for the booklet. They also collaborated, along with Ken Alexander, Jim Williams, Maurice Bruce and other colleagues at the Extra-mural Department, in developing the idea for Northern College. A product of this collaboration is that in North Derbyshire and South Yorkshire trade unionism acquired a more reflective and cerebral quality, which was harnessed to the willingness and the ability to take collective action.

Michael and Eleanor had been members of the Communist Party, but both left following the Khrushchev speech to the 20th Congress of the Communist Party of the Soviet Union, in 1956, and the Red Army crushing of the Hungarian revolution. However, they both remained active and committed socialists. They both became ex-communists but never anti-communists. Michael found himself in the intellectual milieu that led to the creation of the *New Left Review* and served on its editorial board. He helped draft and actively supported the May Day Manifesto of 1966, published by Penguin in 1968. It was, however, his partnership with Ken Coates and Tony Topham, after 1962, which may have had the most influence. It was out of this collaboration that two important initiatives were launched. These were the formation of the Society of Industrial Tutors and the Institute for Workers' Control.

The Society of Industrial Tutors, launched in 1963, was a collegial organisation of industrial studies tutors which promoted the discursive consideration of their professional role. It focused on study issues and avoided terms and conditions issues. Michael was chair and later president of the Society from 1963 to 1983. During those early years, and on behalf of the Society, Michael built up a good working relationship with senior staff in the TUC Education Service, producing a memorable and significant series of TV programmes and books. This project, known as the BBC-TUC-WEA multi-media initiative, made available to tutors and students a core set of texts dealing with a wide range of topics from workplace organisation to work measurement and payment systems. As was suggested by John Holford in his study of the TUC Education Service, this initiative preceded the development of a nationally delivered but consistent model for trade union education by the TUC itself.

The Institute for Workers' Control evolved out of a meeting of adult tutors and ultimately shook the ground of national politics, influencing the Labour Party and the TUC. At its height it convened working conferences of thousands of trade union convenors, shop stewards, active members, and full time officers, including trade union national committee members and general secretaries. Under its aegis, a large number of workplace groups across the country, in a wide range of industries and occupations, produced detailed and costed plans for the transformation of their industries or their companies (see *The Spokesman*, no. 116, pages 47-53, 2012). During the 1960s and 1970s, Michael was heavily involved in the IWC's activities and in writing for its publications. It was perfectly understandable that, with the number of experienced adult tutors among its founders, the IWC would have an approach to its activity which was rooted in the experience of trade union adult learning. This was explored in some detail by Ken Coates in an essay on 'Education as a lifelong experience' (*Workers' Control: Another World Is Possible*, 2003). The approach is to encourage learning from experience as a tool for further development. Of course, this learning is aided by tutorial guidance in the form of lectures, recommended reading, research and discussion, but the overall aim is to equip the learner, singly or in collaboration with others, to discover and to learn from this discovery. This approach seeks to develop self-activating and lifelong students and researchers and to imbue trade union activists with a reflective and thoughtful dimension to their collective activity.

Michael left Northern College in 1984, and proceeded to add to his long list of activities and publications. Most significantly he was the founding chair and director of Third World Information & Twin Trading, 1985-97, and continued as a trustee and director after 1997, a role he maintained until recently. It was natural for Michael to contribute to building an initiative such as this because it combined his interests as an economist and adult educator along with his commitment to others with roots in his socialism. During a tutorial session with NUM day release students in 1965, he raised the question and challenged the students to reflect on 'why was economic growth necessary?' This reflected the attention he was giving to the environmental impact of industrialisation and the unequal distribution of the benefits of economic activity. The ensuing discussion embraced the quite different experiences of living in the 'first' world and living in the 'third' world. The constellation of issues raised by that discussion continue to dominate our lives, probably more so in 2015 that in 1965. This work was supplemented with publications on 'fair trade' and global trading relationships and reprised his much earlier study, *After Imperialism* (1963).

Parallel to his work on fair trade, Michael was also a director of the Bertrand Russell Peace Foundation. The Russell Foundation takes a global approach to its mission and recognises that enduring 'peace' can be built out of global mutuality and respect for each other. The Foundation offers a connection that unites Michael's life's work as a socialist economist and adult educator. A sustainable world requires that we treat each other with fairness and respect, and our daily experience offers the material evidence for our learning.

The Northern College for Residential Adult Education was formally opened for students in Autumn 1978. It has endured for thirty-seven years, against formidable odds (See Ball, M and Hampton, W (2004) *The Northern College*, NIACE). During this time, Newbattle Abbey, Coleg Harlech, the Catholic adult college in Oxford, and the LSE Adult Residential Programme have withdrawn from residential adult education. The Northern College currently has its fourth Principal. However, Michael's task was to build a new college. He recounts the amazing story in his contribution to the 2004 volume dedicated to the success of the first twenty-five years. Each of the four Principals has contributed to the continuing success and reputation of the College, a new college built on the foundations of mutual respect and a faith in the ability of working class adult women and men to refresh their lives with the aid of learning and to refresh their communities and their neighbours' lives. Those values were established in the foundation of the College and find a faithful reflection in the mission of the College in 2015.

Michael Barratt Brown
Select Bibliography

After Imperialism (1963) Heinemann

Essays on Imperialism (1970) Spokesman

From Labourism to Socialism (1972) Spokesman

Economics of Imperialism (1974) Penguin

Resources and the Environment (1976) Spokesman

Information at Work (1978) Heinemann

Models in Political Economy (1984) Penguin

European Union: Fortress or Democracy? (1991) Spokesman

Short Changed: Africa and World Trade with Pauline Tiffen (1992) Pluto

Fair Trade: Reform and Realities in the International Trading System (1993) Zed Press

Africa's Choices: After 30 Years of the World Bank (1995) Penguin

The Blair Revelation: Delivery for Whom? with Ken Coates (1996) Spokesman

An African Road for Development: Are we all Romantics? (1997)

University of Leeds, African Studies Unit

Defending the Welfare State (1998) Spokesman

Young Person's Guide to the Global Crisis and the Alternative (1999) Spokesman

The Captive Party (2001) Spokesman

From Tito to Milosevic: Yugoslavia, the Lost Country (2005) Merlin Press.

Seekers: A Twentieth Century Life, (autobiography, 2013) Spokesman

In addition, numerous articles and reviews are to be found in:

New Left Review

The Spokesman

Oxford University Institute of Statistics Bulletin

Yorkshire Bulletin of Statistics

Review of African Political Economy

Capital and Class

The Industrial Tutor

Institute for Workers' Control Bulletin

See also: Ball, M and Hampton,W. (Eds), (2004) *The Northern College: twenty-five years of adult learning*, NIACE

In the classroom

A Personal Tribute

Victoria Bawtree

At Machu Pichu, Peru

The author co-edited The Post-Development Reader *(Zed, 1997).*

I first met Michael at an international conference on Marxism organised in 1964 by the Gramsci Institute in Rome. As far as I remember, Michael and I were the only English participants in the conference – Marxist studies were only slowly becoming fashionable in England at that time and Michael, as always, was at the forefront of new and independent left-wing thinking. As for me, I came from a very non-political family, but 15 months working in Sicily as a volunteer with Danilo Dolci and two years spent in New York had started to open my eyes, and I drank in all that Michael had to say. At the time I was doing a correspondence course in economics with London University, while earning my iving at the UN Food and Agriculture Organization, and he gave me invaluable advice about books that were not exactly in line with those recommended by my course, for which I shall always be grateful. His *After Imperialism* had come out the year before and I remember how angry he was that the publisher had insisted on removing the question mark after 'imperialism' (in subsequent editions I noticed that the question mark had been re-inserted). This book had a profound influence on me – and no doubt on many others: Ernest Mandel described it, in 1964, as 'undoubtedly one of the most important economic works recently published in English'. And for me, who already had a certain international experiencc in UN circles in New York, I was especially interested in the recently de-colonised countries of the Third World, as it was called at that time. I continued to work at FAO, finally obtaining a B.Sc (Econ.) from London University, but I never returned to live in England. However, I paid frequent visits there and Robin Hood Farm in Derbyshire was a must on my travel itinerary round the country. How many hours did I spend with Michael in his large library, while he would pull out books and say 'Have you read this?' 'You surely know that?' I usually didn't.

After the library session we would descend to the kitchen to join Eleanor – in the first years there was also Eileen, Michael's lovely mother. We always had excellent meals, cooked by both Eleanor and Michael, and they invariably included home-grown fruit and vegetables from the Robin Hood Farm garden lovingly tended by Michael. (I was always impressed by their huge freezer containing scores of little packages of anything from gooseberries to broad beans, all neatly labelled and dated.) And how we talked! Intellectually isolated as I was, living in Italy and not in an academic milieu, it was heady stuff indeed. Nor was it all only about books. Michael had prodigious energy and liked to talk about the things he was always busy organising: if it wasn't helping to set up, with Ken Coates, his very close friend, the Institute for Workers' Control, and the Industrial Tutors' Society, it was founding the Northern College for adult education and, in later years, TWIN and Twin Trading Ltd, promoting the products of small peasant co-operatives in Africa and Central America.

Myself being somewhat of a traveller (in the course of my work), I was always interested to hear of Michael's sojourns abroad, usually to lecture on adult education at universities in countries such as India, Australia, Papua New Guinea, Japan, and as part of medical delegations with Eleanor, to China and Cuba. Always on his visits, Michael would go for his daily jog, as he did on his many visits to me in the south of France, where our rural setting was rather more suitable for running than some of the cities he stayed in, where traffic and pollution detracted from the pleasure. Always, also, Michael wrote a detailed account of what he had seen and heard, whom he had met, and his general impressions of the host country. He used to send copies to his friends and the writing was so vivid that you felt that you had actually been there yourself.

Michael also sent me copies of the many books he was writing, in between the organising, the teaching, the travelling. Several focused on economics, which was his main field and one of his books that I personally found especially useful was his *Models in Political Economy: A Guide to the Arguments*, which was published in 1984. I was delighted to learn, later, that it was known and cited in Italy and France. International trade was his speciality and this of course served as a background for his vigorous activities on behalf of fair trade.

Another of his books that was of special interest to me was *Africa's Choices: After 30 years of the World Bank*, which was published in 1995. In the 1970s and 80s, I had been closely following what had been happening in Africa and in reading the manuscript of this book I was able to make a number of suggestions that he adopted. I had been actively supporting (as far as one can in a UN organization!) the African liberation movements, the anti-apartheid campaigns and, in this connection, Michael introduced me to another good friend of his, Basil Davidson, the British historian on Africa who had been much involved with these movements, especially in the former Portuguese colonies. Subsequently, I worked for a year in Mozambique, in 1985, and a year in liberated Eritrea, in 1996. Like me, Michael became very interested in the heroic struggle of the Eritreans to free themselves from the yoke of neighbouring Ethiopia, which they finally won – only to find that an even greater struggle lay ahead if they were to be free to set their own economic and political policies. As, indeed, has been the case – and still is – for almost all 'decolonised' countries, a situation that Michael always strongly deplored and denounced.

There is so much that could be said to describe Michael's activities during his long life, and this will be better done by people who have worked with him in England. Here I have concentrated more on his international connections because that is what we shared in common. There is no doubt that he had a tremendous influence on what I have done, not least with the FAO publication *Ideas and Action*, which I edited for so many years. When the bureaucrats finally got their way and decided to kill it, Michael was one of many who protested. As he wrote in the last issue, in 1987:

> 'Innovation is a rare bird in any field … At the international level, FAO's bimonthly bulletin *Ideas and Action* would, on any count, have a place high up on the list [of innovations]. From its earliest numbers issued nearly 25 years ago, it has been more than just a bulletin describing any number of interesting experiments in rural development, mainly in the Third World. It has set out to be and has succeeded in being a sharp weapon in the continuing struggle of human beings the world over to free themselves, not only from hunger and ignorance, but from prejudice and apathy.'

If this was in fact the case, Michael Barratt Brown must take much of the credit.

SEEKERS

A Twentieth Century Life

Michael Barratt Brown

Purposeful Life

Stan Newens

I first met Michael Barratt Brown in the mid-1950s, when he lived in a former public house, the 'Live and Let Live' in Dedham, north Essex, near the border with Suffolk. At the time, Michael was working as a WEA lecturer and was a deeply committed socialist with whom I, as a left-wing Labour Party activist in Essex, was anxious to make contact.

In 1958, Michael moved to Sheffield to become a lecturer in Sheffield University's Extra-Mural Department where, along with Royden Harrison, another dedicated socialist, he lectured to miners and other trade unionists. Later he became Principal of the Northern College, which does magnificent work in adult education.

In 1956, Michael left the Communist Party over disillusionment culminating in Soviet intervention in Hungary, and he became a prominent figure in the New Left. He was a member of the editorial board of the *New Reasoner*, founded by two ex-communists, E P Thompson and John Saville, and also of that of the *Universities & Left Review*, founded by left-wing students Raphael Samuel, Stuart Hall, Charles Taylor and Gabriel Pearson. He helped merge the two journals to form *New Left Review*, which is still published. The New Left movement had a seminal effect on the development of left-wing ideas in the 1950s and 60s, and Michael's contribution to it was considerable.

I can remember crowded meetings at which Michael was a key speaker and the brilliance of his political analysis totally captivated his audiences. He buttressed his reputation as a speaker with a steady stream of articles and books. I particularly

Stan Newens was a Labour MP for many years, before his election to the European Parliament in 1984.

◄ *Michael's autobiography* (Spokesman, 2013)

remember an exceptional work, *After Imperialism*, first published in 1963 and subsequently revised and brought up-to-date, but there were many more.

Michael was one of the last survivors of an outstanding generation of socialist intellectuals which included E P Thompson, John Saville, Raphael Samuel, Stuart Hall, Raymond Williams, and many others.

He continued practically up to his death to produce penetrating articles – most of which, in later years, appeared in *The Spokesman*. He was an expert on many topics: former Yugoslavia, where he was in World War Two; British industry and the economy; imperialism and the developing countries, and so on.

Michael will be sadly missed, but his exceptional contribution to socialist thought and the Labour Movement will be remembered. He led a purposeful life and made a rich contribution to left-wing ideas and the wider community.

> 'At the same time that I joined the *New Reasoner* board, I was invited on to the board of *Universities and Left Review*, which had been founded in Oxford among others by Ralph Samuel and Charles Taylor, later a party leader in Canada. We decided to amalgamate the two journals and form *New Left Review* (*NLR*). As a member of both boards, I was made the chairman of the meeting, and I shall not forget the complaints from Ralph Samuel, that he could not take any more "Quaker compromises".'
>
> Michael Barratt Brown, *Seekers*, page 125

Reviews

The Knight versus the Knaves

Sir Michael Barber, *How to Run a Government – so that citizens benefit and taxpayers don't go crazy*, Allen Lane, 2015, 368 pages, ISBN 9780241004975, £16.99

Sir Michael Barber has come a long way since he worked, briefly, as a teacher and thence as head of education at the UK National Union of Teachers. First he was snapped up as an education adviser by Tony Blair, then with New Labour in government, he became head of delivery for just about everything. Indeed he even invented the term 'deliverology' to describe what he does. However, it is in education where Barber has really made his mark. Before I go on to review his latest work therefore, perhaps it would be helpful to give a brief overview of the education movement for which he is one of the most senior power brokers.

The Global Education Reform Movement, or GERM as it is not so affectionately known by teacher activists, has infected education systems, as its name suggests, globally. Its chief advocates are education corporations such as the mighty Pearson (for whom Barber is the chief education adviser), international financial institutions, in particular the World Bank, and neoliberal politicians. Its central narrative is that education is the way to solve the problem of poverty. This conveniently absolves the capitalist system, of which these advocates are some of the primary beneficiaries, of any responsibility for the unequal economic and social relations, which are its true cause. Moreover, it provides ready-made scapegoats for poverty, in the shape of state schools, teachers and, especially, teachers' unions. The putative failure of these groups to 'solve' the problem of poverty drives a series of solutions which make up the architecture of the GERM. These are accountability measures, which demand huge quantities of data, standardised testing and curricula. The data in turn determine such issues as teacher tenure and pay, as well as judgements on the success or failure of state schools. Privatisation, either through vouchers or through chains of private schools, is seen as an ideal way of overcoming state school 'failure'. The purpose of education is framed as the development of human capital and it is to that that the standardised curricula are geared. Teachers lose their professional autonomy and become deliverers of pre-determined content. Teaching unions, which are seen as the main block to the GERM, are to be taken on, sidelined or defeated. Education budgets are cut in line with the requirements of capital for lower taxation. And, of course, corporations such as Pearson, who produce the testing materials, the software, the lesson plans and the teaching materials, as well as chains of private schools, make a vast turnover from an 'industry' which has been identified as 'the last frontier for profit'.

Given Barber's central role in advocating GERM, it comes as no surprise to find that this book takes it as read that the neoliberal playbook of privatisation and austerity is just what the poor need to get out of poverty. He expects to 'see markets becoming steadily more important in meeting the needs and aspirations of citizens'. Prime Ministers are really just like 'CEOs of companies'. Tony Blair's introduction of the market into health and Thatcher's selling off of council houses were all 'truly transformational' (in a good way, of course). Indeed, says Barber, 'the ideological debate should be over by now'. This goes, too, for continuing austerity, or to put it in GERM speak, 'the primacy of outcomes over inputs' or, more prosaically, 'more for less'. Because, as Barber helpfully explains, 'corporate taxation … is becoming ever more difficult to collect as corporations become globalised'. Far from being a problem to be tackled, this is seen as simply a fact, and so the answer is improving the 'productivity of public services'.

It is therefore no surprise to find the book full of praise for a cast of neoliberal pundits and politicians. Words like 'brilliant', 'towering', 'heroic', 'stellar' are generously attributed to such figures as Tony Blair, Andrew Adonis and Arne Duncan, US Secretary of Education; education 'reformers' such as Joel Klein, Paul Vallas and Cami Anderson; and an assortment of political leaders from the global South, such as Sebastian Pinera of Chile or Najib Razac of Malaysia. He is not sparing, either, in praise for himself, quoting a *Times* article, for example, where the journalist says, 'the most important words came not from Tony Blair but from Michael Barber'. To be fair, Barber does also quote one journalist who compares a presentation from him 'as comparable to a lecture from the speaking clock', but fortunately his self-regard is adequate to the task and he takes this 'as confirmation that I was doing my job'. (It is doubtful that any self-respecting teacher would feel the same way, but let that pass.)

This obsessive name-dropping would be mildly amusing or irritating, depending on one's tolerance levels, if it weren't for the fact that Barber makes it clear that he has the ear of most of the movers and shakers he lauds, and there is every reason to believe him. To give one example of his influence, he has been financed, at great expense, by the Department for International Development to 'reform' education in the Pakistani Punjab.

He prescribes the standard GERM remedies: privatisation, vouchers, data collection, performance related pay and tenure, standardised lessons and testing. As usual, his self-belief is, as he might say, stellar; 'the progress has been remarkable'. His enthusiasm is not shared, however, by most of the teachers in the Punjab. There have been many strikes against the 'reforms' and, in particular, against the poverty pay of teachers and the fact that a huge number of them are on even worse-paid temporary contracts. One such strike is ongoing as I write.

This brings me on to consider Barber's attitude to teachers and, in particular, their unions. He divides those involved in education into two groups: producers (teachers and their unions) and consumers (parents and pupils). It is, he says, the consumers who must be considered, not the producers, who tend to have a 'vested interest' and who 'constantly demand increased pay, better conditions, decreased workload or more support'. The idea that all of those things might also be in the interest of the 'consumers' clearly has not occurred to him. Neither has the irony of his attacking teachers' unions for claiming more of the share of taxation for public education, while the corporation he is a highly paid employee of is claiming its share of the public purse, in order to feed its bottom line. If there are vested interests in education, other than those of communities, parents, teachers and pupils, the most egregious ones must surely be those of education corporations.

It would not be an exaggeration to say that this book drips with contempt for teachers. He quotes with horror a Labour politician who says on the radio that 'each teacher knows best in her own classroom', a policy which, he says, has caused the 'plateau' in education in the post-war period. On the other hand, he has nothing but praise for a US school principal who tells her superintendent that he should 'studiously ignore' all complaints for a two month period. He acknowledges that in Finland the policy of 'trust and altruism' towards teachers appears to have worked, but asserts that Finland is 'highly unusual' and quips to governments 'don't try this at home'. At one point he suggests the nuclear option, quoting from the play *Main Street* – 'And you want to reform people like that when dynamite is so cheap?'

If he is contemptuous of teachers in the North, he is downright libellous of many in the global South. He blithely asserts, that 'a quarter of India's teachers do not turn up on any given day'. This is a canard which is frequently quoted in the global media, and one which totally ignores the factors which lead to teacher 'absenteeism'. In South Asia, in so far as it exists, a huge factor is the amount of non-educational tasks such as census taking, organising elections, vaccination campaigns and many more which take teachers out of school, often for several days, and with no substitute teachers when they are out. Moreover, there are no substitute teachers when permanent staff are on maternity leave or ill. Add to that the poverty pay and often appalling conditions in state schools, and the miracle is not that there is some 'absenteeism' but that teachers manage to teach at all. Of the teachers of Ghana, he asserts that the government 'pays teachers well'. He then goes on to slate their competence and commitment on the basis of one anecdote, and states that 'the government is paying for knights and getting knaves'. This insult from a man who, reportedly, was paid £4,000 a day for advising the Punjab government, dismissing Ghana's teachers as knaves, who are on poverty pay and sometimes unpaid for months, and who are working in oversized classes in crumbling schools, is staggering. But then, of course, Barber knows all about the difference between knights and knaves – having himself been knighted by the Queen.

Like all those in the GERM camp, Barber is very clear what constitutes the main 'block' to reform – it is teachers' unions. And in this book he advocates the same tactics as the World Bank – attempt to sideline them if possible, don't allow them to 'drag you into a mud-wrestle in which you and they become equivalent'. Create a 'story' which bigs up the GERM narrative in order to get the community on your side. However, he admits this is not easy. If only communities wouldn't allow themselves to 'become dependent' on public services. If only you could get them to support a 'major government reform'. Alas, he admits, 'you risk being accused of softening up the service for cuts and abdicating responsibility'. Which, of course, is not surprising because, as I have shown above, Barber believes that cuts are necessary and privatisation is the answer.

This is a tedious book and, in many ways, a depressing one, given the degree of influence Barber and education corporations have on school systems globally. However, there is a glimmer of hope. It is not so easy, as Barber admits, to get communities on your side. And despite not having access to the mass media, teachers and their unions have the huge advantage of being embedded in local communities globally and are often some of their most respected members. Together, teachers and their school communities represent a powerful and potentially unstoppable force. This is not the place to detail some of the amazing initiatives by communities, particularly in Latin America, to forge a different kind of education reform, one which takes as its starting point the transformative power of education to create critical and creative thinkers, who can question and change the deeply unjust and unequal economic and social conditions in which we all live and which cause so much misery. But, be sure that even Barber knows that teachers and unions, working with communities, represent the biggest threats to the GERM and similar neoliberal policies.

In a very revealing passage, the author talks of the time when he introduced the so-called 'literacy hour' into English schools, a typical de-professionalising, top-down policy from his unit. He talks about the 'awful feeling' on the morning it was to be introduced:

> 'Would there be a boycott? Would there be demonstrations? Or worse still, what if in spite of all the training that had already taken place, the primary teachers of England just carried on doing what they'd always done? What would government do then? How powerless would I (sic) be? To be ignored: surely a worse fate than to be resisted.'

As it happens, the policy came in and there was no organised resistance. But both Barber's self-identification with government and his fear of union action are clear indications of the battle lines. Increasingly, corporate interests are occupying government at the same time as those resisting are beginning to occupy the squares.

If you, like me, believe that another world is possible and that the biggest 'block' to that world is the interests of global capital, read this book so that you can know your enemy. Or, alternatively, just take my word for it, unless you want a lecture from a particularly self-regarding speaking clock.

Mary Compton

Spotted Owls

Robert W. McChesney, *Blowing the Roof Off the Twenty-First Century: Media, Politics and the Struggle for Post-Capitalist Democracy*, Monthly Review Press, 2014, 272 pages, hardback ISBN 9781583674789, £22

This is a series of essays that have been revised and dressed up as a book and it shows; it is by no means 'seamless'. The first few chapters go over the problems of the US over the last 15-20 years: the wars, the military-industrial-political complex, the corruption, the growing inequality, the marginalisation of social services, the power of corporations and banks, the crash of 2008, etc. There is nothing new here but might be useful to those who missed out on Noam Chomsky, John Pilger, Gore Vidal, William Blum, Mark Curtis et al. I would sooner go back to them; they are better writers.

The author makes clear the US is a democracy in name only, preferring to call it a dollarocracy. He goes into 'economics' in some technical detail, but clearly enough so that even your 'economics-illiterate' reviewer could understand. He often refers to economic 'stagnation' and wonders 'how can capitalists be so short-sighted as to oppose the use of government to build infrastructure, create jobs and end stagnation?' After all, a reasonably vibrant economy puts more money in the population's pockets so they can consume more and add to corporation profits. But no, stagnation and apathy present a lesser threat.

> 'Warren Buffett, one of the three or four wealthiest persons in the world … called attention to the absurd situation that he was paying a lower income tax rate than his employees … "Legislators in Washington," Buffett wrote, "feel compelled to protect" mega-rich people like himself "as if we were spotted owls or some other endangered species".'

That says it all.

Probably the most interesting chapter is the one on America's penal system. Just as its military spending makes up half of the world's military expenditure (the other 50% comes from the spending of all the other armed forces of the world combined), so the USA accounts for almost a quarter of the world's prisoners, yet has only 5% of the world's population. America's gulag is, of course, a form of social control, especially in times of financial crisis. It keeps the poor in their place and it is the poor who make up most of the prison population, the majority black, followed by Latinos, then by whites. I was astonished to learn recently, though this book doesn't mention it, that there are 1,500 on Death Row.

The prison service is lucrative business, too, as it is increasingly privatised. One Virginia mayor is quoted:

> 'I can get $600,000 from the state for a new jail, but I can't get $40 for "Healthy Families",' (a public health programme for infants).

You can skip the interview with the author, a self-serving piece on his involvement with 'alternative media'. It's important, of course, but fairly harmless, otherwise it wouldn't be tolerated. It is a spit in the bucket compared to the onslaught of corporate controlled mass media. Likewise, the self-congratulatory tone in 'The State of Media and Media Reform' in which he states what a brilliant interviewer and interviewee he was.

McChesney's chapter on 'Walter Lippman and the Crisis of Journalism' is more interesting; a brief history of journalism and Walter Lippman is its hero. There is no doubt about it, Lippman was a great journalist, and the author overlooks his 'elitism' because Lippman was an idealist who tried to clean up journalism's act and make it a proper profession:

> 'Reporters needed to create and enforce a professional "code of honour", similar to that adopted and enforced by the professional associations of lawyers and doctors.'

The idealism is undeniable and admirable but, today, seems like pie-in-the-sky.

Part four, 'Bringing it all Back Home', with a long chapter on 'The Cultural Apparatus of Monopoly Capital' is superb. It is the most academic part of the book, guilty of some jargon, so the start is heavy going. But keep at it; it's worth while. For a start, he quotes extensively from C. Wright Mills, E.P. Thompson, Erich Fromm, Raymond Williams and Herbert Marcuse, all brilliant and lucid writers. He is also very good on the decline of the book. I remember, many years ago, being shocked at learning that there were more video shops in the UK than bookshops. Video, of course, is now old technology, along with 78 rpm records, LPs, EPs and cassettes, and the book looks ready to follow the scroll. A note of pessimism enters:

> 'We are back at the state of the Utopians pure and simple; a better world there should be but there ain't no social force in sight to bring it about.' (Paul Baran on Marcuse's *One-Dimensional Man*)

And:

> 'The idea of an alternative culture is radical but limited. It can very easily become a marginal culture; even, at worst, a tolerated play area. It is always certainly inefficient unless it is linked with effective opposition to the dominant system, under which the majority of people are living.' (Raymond Williams)

This is interesting because it contradicts the author's own rather mawkish, upbeat, over-optimistic introduction. There is an air of desperation ('It's not hopeless! No, really, it isn't!'). This is his reaction to the despair he encounters on his lecture tours. He points to the abolition of the slave trade and slavery, and the end of Apartheid. Well, it is true that black Africans are no longer shipped across the ocean and forced to work on cotton and sugar plantations. It is not true that slavery has ended. It is now called 'human trafficking', and it is up there with drugs, arms, oil and pharmaceuticals.

The end of Apartheid has not brought black Africans out of poverty or stopped the violence. South Africa is now ruled by a black corrupt élite which sends in black police to shoot down striking black miners (a black version of Sharpeville).

A little panic-stricken, the author continues:

> 'People rarely accurately predict social change, it is almost always a surprise ... No one saw the Madison uprising coming, or Occupy.'

Rather parochial thinking, don't you think? These were hardly world-shattering events and flopped anyway. The massacre of students in Tiananmen Square, while it got world attention, changed nothing. In both cases, look who is still in charge.

Perhaps the two most unexpected, world-shattering events of our times were, firstly, the take-over of Cuba by a small, battered band of guerrillas. Some 60 years later, after massive repression throughout Latin America and the Caribbean with its dirty wars, invasions and 'disappeared', the Cuban government survives, and there is a resurgence of Bolívar-type resistance throughout the continent; maybe a glimmer of hope there.

Secondly, there was the collapse of the Soviet Union and the fall of the Berlin Wall. Both were hailed as a triumph for 'Western values', a victory for the 'international community', i.e. the countries dominated by the USA; the anti-Christ had been defeated. And what happened? Russia was ruled by gangsters and the life-expectancy of Russians dropped significantly while the West went to permanent war and went bust. We now have round two of the Cold War.

'Ah,' you say, 'but there *is* hope: look at the new Greek government and the rise of Podemos in Spain, that heralds real change.' Maybe, and maybe it will be a change we hadn't contemplated: a return of the Greek colonels and another Spanish civil war?

Nigel Potter

Being St Ann's

Lisa McKenzie, *Getting By: Estates, class and culture in austerity Britain*, Policy Press, 2015, 224 pages, paperback ISBN 9781447309956, £14.99

St Ann's is mainly a 1970s council estate in Nottingham. *Getting By* is based on Lisa McKenzie's experiences of living there and researching what white women with mixed race children think about their lives and this area. More recently, she has added another dimension to her work, including the black men who live there too. McKenzie's work is based on 'new St Ann's', but this area was built on the wreckage of 'old St Ann's', which housed Nottingham's lace workers in the nineteenth century. 'Old St Ann's' was first researched by a team of adult education and university student volunteers led by Ken Coates and Richard Silburn in the late 1960s with the findings eventually published as *Poverty: the Forgotten Englishmen* (1970). Lisa McKenzie read that book and decided that the time was right to find out how the area's current residents felt about themselves and 'new St Ann's'.

In 1966, visiting St Ann's was described by Coates and Silburn as akin to stepping back into the pages of Dickens or Gissing. As they put it, 'long black slum terraces crawled along cobbled streets' with thousands of mainly two-bedroom Victorian houses crammed into a two mile area, close to the city centre. Most had outside toilets, no hot water and no bathrooms. Rubbish, large numbers of dogs and rats were commonplace. Cracked sewers and leaking roofs, thin walls (allowing little privacy) and rotting floors were not unusual. Originally built for Nottingham's industrial white working-class, by the late 1960s it also had become home to Italians, Poles, Jamaicans, Pakistanis, Irish, Scottish and Welsh people, who moved here to find work. Pubs abounded and corner shops were 'open all hours', including those catering for Italians, Poles, West Indians and Chinese. Open-air markets sold anything. Pawnshops bought anything, even pawn tickets. But the council had started seeing St Ann's as an area where they could re-house 'problem families'. 'Vagrants' had invited themselves in, too, dossing in the many houses left empty as they became literally uninhabitable. The future of 'old St Ann's' was becoming increasingly uncertain.

'New St Ann's' is the council estate that replaced the 30,000 houses, pubs and shops that disappeared when 'old St Ann's' was demolished as part of Nottingham City Council's slum clearance programme in the early 1970s. The Wimpey-built mainly council-owned housing is much better quality now than the old slums, though only 14,000 people live in 'new St Ann's'. Over the years, residents' initial pride in their gleaming homes and the area's modern image has diminished. Some addresses are hard to find, even for delivery van drivers and taxi firms. This is not surprising given the design of the estate, which McKenzie describes as

> '... rows and rows of the same ... grey concrete pebbledash houses, flats and two-storey maisonettes ... suspended on footbridges on several levels, [which] reach each other through underpasses and subways, which are often very dark and covered in rubbish and debris.'

The St Ann's Well Road, the main road running across the estate, was a 'vibrant shopping centre' in 'old St Ann's'. It is now a busy but 'soulless' commuter road from the city centre out towards the suburbs. There are few local shops and deteriorating community facilities. The Co-op and the community laundrette closed in 2012. The Co-op's premises are empty: the laundrette has become St Ann's food bank. The last pub in St Ann's closed its doors in 2013. If this was not starkly depressing enough, the area's reputation in the local and national media has been troubled. There are high levels of unemployment and poverty, but 'new St Ann's' is also 'known' for crime, drug-dealing, prostitution and gang violence, and media stories regularly appear which seem to confirm this picture.

That said, St Ann's is still the first place in Nottingham where many newcomers come looking for a home or work (or both). They may settle for a while, if not permanently. Fifty years on, the newcomers are from Eastern Europe, Somalia and the Middle East, but varying degrees of welcome and friction between old and new residents and different ethnic groups remains. Nearly three-quarters of school pupils in the area have a minority ethnic background (African, Afro-Caribbean, small numbers of Asian and Chinese). Mixed-race pupils are in the majority.

In McKenzie's view, in 2014, St Ann's is 'as much about being Jamaican as it is about the white working-class'. Many long-established West Indian families are highly respected because their family connections reach back to the 1950s and 1960s and 'old St Ann's'. Local white women and black Jamaican men have married or lived together since then. There are now many white women bringing up mixed-race children in St Ann's. Many Jamaican men have become husbands, partners or, as McKenzie calls them, 'babyfathers'. What did they think about themselves, their children and the area?

McKenzie found that most of the women wanted to talk about how St Ann's was viewed by outsiders, but many also commented on how they were seen, too, as women living in the area. Louise's Irish family came to St Ann's in the 1950s when she was a child. In 1970, they were one of the first to be re-housed in a 'new St Ann's' council house. They believed their new home was 'posh', but over the years their attitude has changed. Louise commented: 'when you tell people where you come from, yeah, you feel like you know that they like class you like rough and ready'. Partly, this is a reflection on the area and its reputation (which affects all women living in St Ann's), but this is also linked to the way women with mixed-race children are viewed. Not only are they working-class white women (which currently de-feminises them into 'rough' women), they are also women who have had sexual relations with black men. Because of this association, evidenced by their mixed-race children, they are regarded as 'slags' or 'prostitutes' by outsiders, 'tainted' or lacking in morals. This is not new. Shirley, whose eldest daughter is now 30, said that when she first started dating black men, she was told 'no white man will touch you now'.

These destructive attitudes never surface within St Ann's, a place where women with mixed-race children are accepted as 'normal'. St Ann's is 'home, community and also their place of safety'. It is also where they feel valued as mothers. Many of the women talked about 'being St Ann's' (in relation to the place and the people) and how important it is to them. Finding personal value on the estate is about 'mixing' but only in certain ways. Currently, McKenzie has identified 'insiders' (who are probably mainly white but could belong to any ethnic group), Jamaicans (who are regarded as a valuable resource), and 'outsiders' who are 'culturally foreign'. 'Outsiders' include Iraqis, some Eastern Europeans and Africans and, while new boundaries are being erected against them (sometimes with actual physical violence), the boundaries between the white community and Jamaican community have long gone. But it is the culture of black Jamaican masculinity that is particularly valued.

'Mixing' has been occurring for fifty years between the white (including Irish) and Jamaican families. A 'hybrid' or 'interchangeable' culture is being created, which itself is constantly changing. The 'hybrid' culture in evidence now includes ways of speaking (Jamaican words), dressing with style (gold jewellery and expensive designer clothes), black hairstyles, taste in music, how you cook and what you eat (especially rice 'n' peas 'n' chicken). Mixed-race children are loved and valued, not just as beautiful children, but also because they are an indication of 'mixing' – they are 'multi-cultural' in a way that white children can never be.

When women moved outside the area, they felt differently about themselves, especially when they deal with statutory organisations. McKenzie thought a woman's respectability was the first casualty in these exchanges. Gina said she did not usually feel 'poor' or 'excluded', but when she came into contact with officials who wanted her address and personal details, she felt that her status as a lone parent, her mixed-race children, and her home in St Ann's were weighed up, in what was often a pause. 'I know what they're thinking you can see it ticking over in their brain as you wait for them to think "oh, it's one of them from there".' Other women commented that they were 'looked down on' and consequently 'never felt good enough'. As McKenzie comments, this often unavoidable contact is hurtful, damaging and disrespectful. No wonder St Ann's itself is regarded by these women as a safe place for them.

Nearly half of the working-age population in St Ann's is unemployed with young mixed-race and African-Caribbean men and women particularly likely to be without formal employment. McKenzie found that many of the men she spoke to were literally off every agency's radar as they had no official addresses and avoided claiming benefits, preferring to work part-time or have 'little runnings' (a Jamaican word for dealings or business). Alternatively, they might 'hussle', selling drugs, cannabis, steroids, Viagra, cocaine, crack or heroin and sometimes stolen goods.

'Business' was important to all of them, including 'chatting business', regularly talking to each other on their Blackberries to find out what was happening to friends, family and a myriad of contacts on the estate. They often 'passed by' their partner's, girlfriend's or mother's homes, always checking before they arrived (to avoid officialdom) and never staying long. 'Passing by' like this 'meant a lifestyle and a transient identity on the estate for the men', but it was also intrinsically linked to their different understanding of 'being St Ann's'. It was about reducing risk and maintaining territory.

McKenzie found that getting a job (assuming one was available) might be a problem for these men. If it was low-paid and low-skilled, it might reduce their status and the respect in which they were held by friends and family. This carried too much risk and potential loss to be contemplated. Typical of many of the men McKenzie spoke to about getting a job was Dread. He is a middle-aged African Caribbean man who divides his time between his own flat and that of his partner. He commented:

> 'There's no jobs here for anyone, what can I do now, I used to work for the council as a gardener, I liked that but that's gone now, I'm not doing no gay job in a call centre.'

His partner's 19-year-old son, Raphel, sold drugs. Although potentially lucrative, Dread could see no point in Raphel feuding over drug territory outside St Ann's. It was dangerous enough within the estate. His advice to him weighed up losses and risks:

> '… there's enough crackheads for all of you to sell drugs to them, let's be honest about it here … there's enough crackheads for all of you to make money without dying, let's be honest … killing each other doesn't make sense, life's hard enough here, just do your business and done.'

Della was just relieved that her son was still alive and making some money.

Most of the women were 'extremely angry and worried' about the amount of drug-dealing and drug use on the estate, which had been growing for over twenty years and which they felt was made worse by the number of dark alleyways and layout of the estate itself. At the same time, McKenzie shows that there is

> 'a kind of acceptance that it was … another part of the fabric of the estate, and many of the women had involvements weak and strong in the drug economy in

> St Ann's because of friends, family members, neighbours and sometimes even their own involvement.'

It is important to say that this is not unique to St Ann's, but McKenzie indicates just how commonplace drug use and drug dealing have become in the area if you know what you are looking for.

There is much more in *Getting By*, but space precludes a more detailed review. McKenzie has, on the whole, managed to transform several academic pieces of work, including her doctorate, into an accessible book full of humanity and honesty about St Ann's and some of the people who live there. This is no mean feat: writing for a new and potentially non-academic audience is hard. For this reason, it might have been better to include more commentary on the significance of Pierre Bourdieu for the uninitiated. The proofreading could have been stronger, too, in places ('deficit' or 'deficient'?) But these suggestions are for the future: what this book says about St Ann's is far more important.

Looking at 'old St Ann's' (in Coates and Silburn) and 'new St Ann's' (in McKenzie) it is possible to see what has been gained and lost. Better housing (albeit with poor layouts) has replaced slums, but community organisations, markets and shops have disappeared. A new hybrid culture has grown between the white and Jamaican communities, replacing old antagonisms, but new ones are surfacing between established residents and newcomers. The drug economy (and associated violence) has grown, too. The residents interviewed by McKenzie have been 'getting by' despite all this. But some have gone to prison and some have died. People at the end of their tethers, in a corner with little or nothing to hope for (not even family life), can explode in sudden anger, as she points out. In the last five years, Cameron and company have pushed people and created places that are now close to the edge, including St Ann's. Where do we go from here? Unfortunately, this is anyone's guess.

Cathy Davis

Ferocious Love

Naomi Klein, *This Changes Everything: Capitalism vs. Climate*, 576 pages, Allen Lane, hardback ISBN 9781846145056, £20.00, Penguin, paperback ISBN 9780241956182, £8.99

In the context of *This Changes Everything*, there are several ways of looking at climate change. There are the straightforward deniers; and then those who say the problem can be solved by the ingenuity of humankind. In the chapter on 'Green Billionaires', for example, two such techniques are mentioned: a machine for the physical extraction of carbon dioxide from the atmosphere, and a device for the deflection of the sun's rays. Finally, there are those who, like the author, think there is just enough time, with the available sustainable technology, to hold warming below 2° centigrade, with the small proviso that we abolish capitalism, certainly in its present neoliberal manifestation. Naomi Klein has written a book that she hopes will be a clarion call to all the progressive forces, trade unionists, community activists, socialists and environmentalists to converge in this effort, not just to save the planet but to transform our relationship with each other and with the planet. If you find the idea daunting then, unfortunately, there is the added problem that we haven't got all that long to achieve the necessary transformation.

This Changes Everything should, however, excite our optimism as, apart from the sheer logic of its argument, the author has spent the last six years researching and visiting people at the sharp end of the struggle to halt climate change. Her anecdotal comments serve to show the extent of protest around the globe, where thousands, possibly millions, are on the march against the activities of the fossil fuel corporate behemoths. For example, there is a momentous battle raging across Canada and the USA to halt the Keystone XL Pipeline System, which is planned to transport highly corrosive tar sands across a swathe of the North American continent. The Republican dominated Senate is trying to overturn the veto that the equivocating Obama finally placed on the project.

Put simply, Klein's thesis is that after wasting most of the years from the late 1980s to the present, through governmental inertia and failure to agree, a period when more modest adjustments could have been made, we are now faced with just enough time to halt a global temperature rise of 2° centigrade if the following steps are taken. This has been called the 'procrastination penalty' for now a radical response is required necessitating the complete abandonment of neoliberalism, in fact the abolition of capitalism itself. To prevent global climate catastrophe, measures need to be taken that are anathema to neoliberal capitalism. We have to have a world where small-scale planned resources are democratically decided, where transport is communal not individual, where rampant inequality is curtailed by wealth redistribution, where agriculture is designed to feed *all* the world's population, where the pursuit of growth and profit are replaced by ecological virtues, and where humanistic, altruistic values replace those of the dominating competitive ethos of the market. Klein calls it 'ferocious love', and calls for the 'regeneration of society'. No wonder Lord Lawson says 'Green is the new Red'. It has to be a society that cares deeply about the shape of the future we leave for posterity, and that treats the Earth with respect and not as a storehouse to plunder. Pretty heady stuff this, but the author thinks that millions of people are near to breaking with this riven world of ours.

There are two principal factors here: is Man capable of such an altruistic endeavour and can enough people be convinced that climate change is such a pressing danger? On the first point, Klein quotes the experience of wartime, particularly during World War Two; but, whilst there were many remarkable examples of collective solidarity and bravery, was it more the result of the wartime state monopoly of the media, the state welding the national effort, and the brutal nature and immediacy of the enemy? Climate change, at least in the advanced nations, is not likely to have the same impact on the popular mind as a world war and, while parts of Africa and the Middle East are enveloped in bloody chaos, we have seen the hostility there can be to migrants, across Europe. The present ruling élite in the developed world is mesmerised by markets as an automatic solution, and can see no other way forward. They establish international conferences, committees and investigatory bodies whose final statements may dominate the media for a few days, but show few signs of creating the urgency needed in the corridors of power. Their much vaunted market mechanisms for dealing with emissions, (the carbon offset and trading markets), turned into a flop as far as reducing emissions, but at the same time made a lot of money for some investors in highly dubious ventures. So it would take a major miracle for the political and economic élites to countenance the freeing up of the media to explain the dramatic nature of the danger from climate change. The whole of the neoliberal philosophy and its attendant institutions are lined up against the sort of radical solution the author has in mind.

It is, however, those same advanced nations who would have to take the unpleasant medicine first. The reason being that they are amongst the highest emitters but possess the civil infrastructure to cope with a sudden drop in power output. Secondly, and perhaps more importantly, by shutting off the major part of these emissions there would be a massive immediate effect, hopefully preventing reaching the 2° increase ceiling. Neither is it the case that in Klein's schema the populace of the developed nations would have its standard of living cut dramatically. It probably means a standard of living that needs to fall to the equivalent norms of the 1970s; not too hard a time for those of us who lived through that period and, of course, hopefully, many things would be better. However, it is going to take an enormous effort to get people in the advanced European countries to think in the radical terms necessary and come to equate the vagaries of the present day weather with a more generalised threat of catastrophic climate change. In this context the International Energy Agency has warned that if, by 2017, steps have not been taken to bring emissions under control then the prospects for life on Earth are bleak indeed.

Naomi Klein's best example of the correct steps being taken is the small-scale energy developments where local groups take the grassroots initiatives. In Germany, local authorities in Munich and Hamburg have made a stand against privatisation, voting to instigate their own brand of 'municipal socialism' by not allowing the utilities to own their gas, electricity and communal heating systems, keeping it in civic ownership. There has been a very significant boost to renewable energy in Germany, as well as the phasing out of nuclear power. Unbelievably, green energy programmes have even been targeted by the World Trade Organization for being protectionist and cosseting programmes for reasons of national trade; there have also been projects attacked by the WTO in Klein's home country, Canada; China filed a complaint over wind power schemes in Italy and Greece; and India has also fallen foul of WTO rules for the same reason.

This Changes Everything has exhaustive analysis of the climate-change-deniers and their association with corporate wealth, in particular the wealth of the big emitters: the oil, mining and transport industries. There are billionaires who give generously to institutions pouring out climate change denial propaganda and they are also mentioned, but it is a sad fact that whilst these deniers have been comprehensively defeated in the scientific argument, their ideology of greed and selfishness is still euphemistically sanctified in the everyday life of commerce. Then, of course, there are the 'green billionaires', Branson and Gates for example, who encourage 'geo-engineering', deflectors in space, carbon capture machines, or the fall-back position of colonising Mars if things get too hot on Earth. Klein thinks we can dismiss these as storylines that befit the latest sci-fi Hollywood blockbuster.

In two books, *No Logo* and *The Shock Doctrine*, Naomi Klein had established herself globally as a powerful international critic of neoliberal economics and power, and the consumerist culture. She makes the case that the pressure for social change necessitates a conjoining of existing movements and that climate change is the way to make the connections between North and South, from the anti-fracking protestors and the landless workers kicked off their land by the encroachment of transnational agribusiness to the localised power communities taking back resources from privatisation. The movement should be about a 'global push' for liberation, equality and community involvement as the only mechanisms for rescuing us from disastrous climate change. Within civil society the division between 'activists' and 'regular people' should be broken down in communities under threat, the native North Americans being an example. This movement will not win the argument by contrasting the costs of sustainable schemes with those of marketised carbon counting: the movement needs to take the moral high ground and show that we want a better world for our children and we do not want to bequeath them a poisoned chalice. No doubt there will be another crisis, which may bring us all out on the streets again, but this time, in the words of the author –

> 'It must be the catalyst to actually build a world that will keep us safe. The stakes are simply too high, and time too short to settle for anything less.'

John Daniels

Grand Tourer

Edited by Glenn Ligon, with Alex Farquharson and Francesco Manacorda, *Glenn Ligon: Encounters and Collisions*, Nottingham Contemporary and Tate, 2015, 284 pages, hardback ISBN 9781849763561, £24.99

Where once there was 'Bear', Steve McQueen's muscular video, now hangs Franz Hals' 'Portrait of a Man'. During summer 2015, it's as though Nottingham Contemporary is sub-let to the fabulously wealthy Stoker Cavendish, Duke of Devonshire and denizen of Chatsworth. Where, during the chilly spring, Glenn Ligon challenged us with his self-curated 'museum' of intimacy, connection and black warmth, now, in the dog days of July and August, we are translated to the pannelled luxury 'goods' purloined, purchased or otherwise obtained by 'well-off' scions of the House of Cavendish. Pablo Bronstein, with an auctioneer's eye, has assembled some 60 pieces of 'incalculable' value. Such treasures sit amidst Bronstein's own Chatsworth-in-Space, *Star Wars* like, and along his *Via Appia*, conjuring Shelley's *Ozymandias*:

'Look upon my works, ye Mighty, and despair!'

Juxtaposition is a hallmark of the Contemporary's 'challenging' programme, as Glenn Ligon called it when recounting his response to

Collection of halley k harrisburg and Michael Rosenfeld

director Alex Farquharson's invitation to curate what became his stunning show, *Encounters and Collisions*, now transferred to Tate Liverpool. Previously, two of the Contemporary's four galleries featured one show, such as Thomas Demand's large-scale photographs, in 2012, while galleries three and four, which look outside onto High Pavement, were given over to *Decolonising Architecture Art Residency*, which liberated Palestine in the heart of England.

This year, juxtaposition is more temporal than spatial. Time and again, one was drawn back to *Encounters and Collisions* and the centrality of James Baldwin's presence on the edge of figurative and abstract in Beauford Delaney's scintillating portrait (above), across the room from the coal dust, oil and polymer of 'Stranger 23', Glenn Ligon's sparkling, large-scale black surface, which takes its title from Baldwin's luminous writing. *The Grand Tour*, which has brought Chatsworth to town, has a more centrifugal intent. Candidly, it is about bringing paying visitors to Notts and Derbyshire, and culture is the bait, as Stoker told us on opening day.

Glenn Ligon: Encounters and Collisions, the book as part of the show, maintains the sky-high standard of Nottingham Contemporary publications. Structured around Ligon's letters to artists, living and dead, whose works interact with Ligon's own in the show, it is an accessible and engaging adventure in high quality lithography. Farquharson has a real talent for such publishing, as revealed in *Aquatopia*, the blockbuster splash of summer 2013 (see *Spokesman 122*).

Tony Simpson

www.nottinghamcontemporary.org

Glenn Ligon

Encounters and Collisions

ONE FOR ALL...
...AND ALL F*** GREECE!
© Steve Bell 2015 - 3841 · 30·6·15